PRAISE FOR DR. JOE

#1 NATIONAL BESTSELLING AUTHOR

"Whether he's celebrating what used to be called 'better living through chemistry' (in DuPont ad slogans, anyway) or exposing the endless fraudulent claims of opportunists, naifs, incompetent researchers and snake-oil salesman (see 'snake oil,' pp. 137-39), plain Joe Schwarcz is unfailingly a clear, engaging and convincing guide."
—*The Globe and Mail*

"There is a vivacity in pure science that somehow gets lost. . . . Schwarcz is doing his part to bring the joie de vivre back."
—*Kansas City Star*

"Schwarcz takes a little history, adds a dash of chemistry and produces a gem." —*Time*

"Falling in love, we all know, is a matter of chemistry. Schwarcz gets his chemistry right, and hooks his readers."
—John C. Polanyi, Nobel Laureate

PRAISE FOR *BRAIN FUEL*

"Packed with scientific questions and answers you didn't even know you had." —*Chatelaine*

"Informative . . . fascinating." —*The Globe and Mail*

DR. JOE'S SCIENCE, SENSE AND NONSENSE

61 nourishing, healthy, bunk-free commentaries
on the chemistry that affects us all

JOE SCHWARCZ, PhD

Director, McGill University
Office for Science and Society

ANCHOR CANADA

Library and Archives Canada Cataloguing in Publication has been applied for.

ISBN 978-0-385-66605-3

Book design: Leah Springate
Printed and bound in the USA

Published in Canada by
Anchor Canada, a division of
Random House of Canada Limited

Visit Random House of Canada Limited's website: www.randomhouse.ca

10   9   8   7   6   5   4   3   2   1

CONTENTS

INTRODUCTION

For me, it was an absolutely stunning remark: "I trust cows more than chemists." And no, these penetrating words did not come from the mouth of some comedian looking for a cheap laugh, but from the lips of an emeritus professor of nutrition and education at Teachers College, Columbia University—not exactly a shabby institution.

Professor Joan Gussow expressed her distrust of chemists when asked whether she favoured margarine or butter. Her choice was butter, she said, because unlike margarine, chemists played no role in its production. I assume she was concerned about the trans fats introduced by the hydrogenation process. Actually, even with those culprits, margarine contains less artery-damaging fat than butter.

Gussow was suggesting that when chemists get involved, nasty things happen. On occasion that may be true, but when in a contest between the benefits and detriments resulting from chemical manipulation—well, there *is* no contest. DuPont's original slogan, "Better things for better living . . . through chemistry," is as true today as it was when introduced in 1935.

In the 1960s and '70s, though, the public's view of chemistry somehow changed. Chemistry went from being a heroic science that furnished us with new medicines, fibres and plastics, to one associated with napalm, Agent Orange and pollution. "Chemical" became a dirty word, with some chemists even suggesting that when communicating with the public the term should be changed to "substance." By 1982 the image of chemistry had been so tainted that DuPont felt the need to drop the words "through chemistry" from its slogan. And this just a brief eighteen years after the company's pavilion at the New York World's Fair in 1964 featured a Broadway-style musical entitled *The Wonderful World of Chemistry.*

It was a great show. I know—I saw it several times, which wasn't that difficult given that it was performed forty-eight times a day!

Everything in the DuPont theatre was made of some newly invented material. Doors featured alkyd resin paint and polyacetal doorknobs, the ceiling was made of polyvinyl fluoride, floors were carpeted with nylon and seats covered with polyvinyl chloride. The polyester curtain rose to reveal dancers in colourful spandex costumes, tapping their polyurethane shoes on an acrylic-glossed stage. There were demonstrations of no-drip paint, colour-changing dyes and nylon fibres being pulled out of test tubes.

Even a vendor outside the pavilion was caught up in the spirit of chemistry. Standing on a soapbox, he expounded on the wonders of high-carbon steel knives that would never lose their shine or their edge because of their chromium, vanadium and molybdenum content. He was quite a showman, as I recall, using one of the miracle knives to slice effortlessly through a whole pack of playing cards. I didn't understand the chemistry at the time, but I was impressed enough to buy the knife, which I still own. Although I have not found the need to cut a pack of cards in half, the knife is a great reminder of chemical ingenuity. And maybe it can even serve as a springboard here to restore some of the shine to the tarnished image of chemistry.

So what do chemists do? As the classic definition goes, they deal with matter and the changes it undergoes. And since anything that occupies space and has mass is matter, chemists are interested in everything. They are interested in the structure of molecules and how this relates to their properties. They are interested in determining which chemicals are found where. But above all, they are interested in manipulating molecules to develop new, useful materials.

Want to know what antioxidants are present in an apple? Or if it is tainted by pesticide residues? Want to know if there is lead in your drinking water? Estrogenic ingredients in your cosmetics? Flame retardants in your blood? Plasticizers in your microwaved foods? If there is a difference in lycopene content between conventional and organic tomatoes? A chemist will have the answers. Do you want to know if an unknown white powder contains anthrax spores, or if an athlete has remnants of drugs in her urine, or if a criminal has traces of gunpowder on his hands? Call a chemist.

Analytical chemists are truly amazing. Imagine being able to detect the presence of a chemical at a concentration of a part per trillion. That's one drop of alcohol in enough water to fill a string of railroad tank cars stretching for ten miles. No less impressive are the chemists who develop batteries, glues, inks, water filters, paper and plastic recycling processes, silicon chips for computers, smoke and carbon monoxide detectors and rocket fuels. And ways to replace those trans fats in foods that concern Professor Gussow and others.

Then there are the chemists who specialize in synthetics. Medicines, dyes, detergents, fire retardants, novel refrigerants and additives for increased gasoline efficiency are in their domain. The ingenuity of chemists working with polymers has given us plastics for blood bags, contact lenses, compact discs, kitchen counters, solar panels and artificial heart valves, as well as fire-resistant suits for firefighters and life-saving bulletproof vests for police officers.

The contributions that chemists have made to the quality of our life are astounding. But it would be naive to think that they come at no cost. There is no free lunch. We have concerns about plasticizers showing up in our blood, dioxins in our air, chlorination by-products in our water and pesticide residues in our food. But it is the development of sophisticated chemical techniques that has made us aware of these issues, and if solutions are to be found, they will come from chemists. And it is worth repeating that whatever problems have been introduced by chemical manipulation, they are far smaller than those that have been solved.

In this book we will explore the role that chemistry plays in our lives, meandering through its history and its connections to our food and health, and poking into some of the strange ideas that people have formulated based on their misunderstandings of the subject. I will try to paint a realistic, unbiased picture of what molecules can and cannot do, separating sense from nonsense in the process. And if I haven't won your trust by the end—consult a cow.

NUTRITION
ISSUES

RADICAL THINKER

ANTIOXIDANTS ARE MAGICAL. At least when it comes to marketing. Just slap the word on the label of a food or beverage and watch sales zoom. That's because even people who have no idea what antioxidants are want more of these substances in their life than less. And they could be right. Or not. It seems the "free radical theory" of disease and aging may not be on as firm a ground as we have been led to believe. And if that is the case, antioxidants may not live up to their exalted status as the key to good health and longevity.

Back in the 1950s, Dr. Denham Harman proposed a theory that at the time seemed rather radical. Many ailments, including cancer and heart disease, as well as the aging process itself, he suggested, were the result of cumulative damage caused by reactive molecular species called free radicals. Since these were by-products of metabolic reactions involved in energy production, their formation in the body was inevitable. Why? Because nutrients derived from food are slowly combusted in mitochondria—the small, membrane-enclosed regions of cells. And, as in any combustion process, oxygen is required. Unfortunately, as oxygen reacts with nutrients to produce energy, it also unleashes some "friendly fire" in the form of the notorious free radicals.

Different types of free radicals can appear, but they all descend from a highly reactive species of oxygen known as superoxide. Radicals are hungry for electrons and try to satisfy their appetites by feeding on innocent molecular bystanders. Since electrons are the glue that binds atoms in a molecule together, molecules that become the targets of free radical attack tend to fragment. Such damage in turn translates to disease, particularly when the victims of free radical onslaught are proteins, fats or molecules of DNA. Harman hypothesized that our bodies deal with free radicals by mounting "antioxidant defences." Vitamins E and C, along with enzymes such as superoxide dismutase, catalase and glutathione peroxidase, were quickly labelled as antioxidants, acknowledging their ability to neutralize free radicals in the test tube. Harman then buttressed his theory by feeding antioxidants to mice, claiming that the animals lived longer. The free radical theory was off and running.

Over the next couple of decades, researchers tested more and more substances in the laboratory for free radical–neutralizing effects, discovering a plethora of antioxidants. The likes of polyphenols, carotenoids and lipoic acid, all found in fruits and vegetables, obliterated free radicals in laboratory experiments. Since populations consuming more fruits and vegetables were known to be healthier, a seemingly obvious explanation now emerged: antioxidants in food prevent disease! A corollary was that dietary antioxidant supplements should also prevent illness, an idea that gave rise to a new market trend. Touting their antioxidant potential, vitamins, minerals, various seed and bark extracts, teas and exotic fruit juices began to vie for the public's attention. And they did so successfully. Sales of antioxidant supplements skyrocketed. Even skin creams joined in the game, hyping the efficacy of their antioxidant ingredients in the battle against the ravages of age.

But now the wheels on the antioxidant bandwagon are developing some squeaks. Since the 1990s, numerous randomized, placebo-controlled trials have investigated the effects of vitamin C, vitamin E, selenium and beta carotene, the classic antioxidants, on cancer and heart disease. While some studies offered hope, the majority failed to show any benefit associated with antioxidant supplements. Some, particularly those using beta carotene, actually suggested potential harm, for smokers at least. Recent studies have not brightened the outlook. The Selenium and Vitamin E Cancer Prevention Trial (SELECT) randomly assigned more than thirty-five thousand men to take either selenium, vitamin E, both or a placebo on a daily basis. The trial was stopped after five and a half years because no differences were observed between the groups in relation to prostate cancer risk. Similar results were noted in the Physicians' Health Study II. About fifteen thousand male physicians were enrolled and asked to take either vitamin C, vitamin E or a placebo for eight years. Neither vitamin had an effect on prostate cancer or total cancer. And now a major study on heart disease has thrown more cold water on the antioxidant theory. This time, the subjects were diabetics, who are, in general, predisposed to heart disease. In the Prevention of Progression of Arterial Disease and Diabetes Trial, more than a thousand adults aged forty or older were assigned, on a random basis, to take either a daily capsule containing a mix of antioxidants or a placebo for eight years. The antioxidants offered no protection against cardiovascular disease.

Could the free radical theory of disease and aging be totally wrong? Not likely. But neither is it the complete answer to the complexity of aging and disease. This is now underlined by a fascinating study on nematode worms at University College in London and Ghent University in Belgium. Granted, people aren't worms (in most cases anyways), but these tiny creatures do serve as an excellent model for the study of aging. Their normal life expectancy is only a few days, so any change is readily noted. The researchers managed to inactivate the worms' genes that code for superoxide dismutase, one of the prime antioxidant enzymes. They expected to see a reduction in life expectancy in response to increased oxidative stress; while there was evidence of increased free radical activity, the worms did not die any sooner. Aging apparently was unaffected by reducing antioxidant activity.

What does all of this mean? That when it comes to health, disease and aging, nothing is as simple as it seems. And so it is with the free radical theory. Evidence is mounting against antioxidants being as important as we thought. But at the same time, we have more and more evidence for the benefits of increased fruit and vegetable intake. A conundrum? Not necessarily. Fruits and vegetables contain hundreds of compounds that may have positive effects that are not yet fully understood, and perhaps we have been overzealous in attributing magical effects to antioxidants.

THE CHEEZ WHIZ EFFECT

NEWS FLASH! The Food and Drug Administration in the U.S. has approved the addition of trans fats to dairy products, meal-replacement bars, soy milk and fruit juice. Now, I know what you're thinking: the agency has gone mad. Either that or it has capitulated to big business. After all, aren't trans fats a nutritional pariah? Are these not the nasty, artery-clogging substances that food producers and restaurants are being urged to purge from their repertoires? Well, the fact is that not all trans fats are fiends. The ones producers are considering adding to foods are not the same as the ones that are terrifying people with their artery-clogging potential. In fact, these trans fats may be good for us. It all comes down to some subtle differences in molecular structure.

Fats are composed of long chains of carbon atoms, joined to each other either by single or double bonds. It is the latter that give rise to the trans conundrum. Depending on the particular arrangement of atoms around the double bond, the chain of carbon atoms will be straight (trans arrangement) or bent (cis arrangement) at the position of the double bond. This may not seem to be very important, but the way a fat engages in biochemical reactions depends on its molecular shape. As a further subtlety, fat molecules can have more than one double bond, each with the cis or trans configuration. Furthermore, the double bonds may be located at different positions along the chain of carbon atoms. The trans double bonds that we worry about are the ones that are separated from each other by more than one single bond. These arise as a result of the hydrogenation process used to harden liquid vegetable oils to improve their keeping properties and to make them more suitable for frying. Unfortunately, these fats don't have excellent properties for keeping our health. They're the trans fats that have been linked with heart disease.

However, trans fats also occur in nature, particularly in meat and dairy products. But in these, the double bonds are separated from each other by just one single bond. Such fats are referred to as "conjugated" and have a completely different health profile from the ones that result from hydrogenation. Conjugated linoleic acid (CLA) contains two double bonds, either of which can assume the cis or trans configuration. The molecules of interest in terms of health benefits are those that have one cis and one trans double bond. Although technically these are trans fats, there is a world of difference between their biological activity and those of the trans fats that may lurk in your doughnut or order of french fries.

The story of the "good" trans fats started with an investigation of the "bad" properties of hamburgers. In the late 1970s, Dr. Michael Pariza at the University of Wisconsin became interested in the chemical reactions that take place when meat is cooked.

Suspicion had been raised that carcinogens may form at high temperatures, and indeed these fears were realized. But much to Pariza's surprise, he also found that cooked hamburger contained some compounds with decided anti-cancer effects. These turned out to be the CLAs, which then understandably excited a number of researchers.

Before long, CLAs were also found to be present in dairy products, originating from chemical reactions in the stomachs of ruminating animals. That's where enzymes convert naturally occurring cis fats in the animals' diet to CLAs. When different dairy products were analyzed for CLA content, researchers were in for a surprise: that curious American concoction known as Cheez Whiz had a higher CLA content than any other food. This finding provided plenty of whimsical fodder for reporters who, with tongues planted firmly in cheeks, began to label Cheez Whiz as the new health food. Of course, the spread is no such thing—its CLA content is easily trumped by unhealthy doses of saturated fat and salt. The Cheez Whiz effect was actually not a boon for CLA research, as critics smirked at the prospect of an ingredient in this odd gustatory creation being touted as potentially "healthy." But they are not smirking at the mention of CLAs now.

A stunning amount of research has been carried out since the initial Cheez Whiz caper brought CLAs into the public spotlight. While no anti-cancer effect has yet been demonstrated in humans, animal models and cell cultures have repeatedly confirmed the initial finding. The greatest excitement, though, has been over CLAs' ability to reduce body fat while enhancing lean body mass. Although supplementing the diet with CLA does not result in a reduction of body weight, it is effective in reducing fat mass and increasing muscle mass. The most significant effects have been seen in people who have lost weight through a low-calorie diet and then put the weight back again, as commonly happens. But subjects who were supplemented with about three grams of CLA a day were more likely to regain the weight as muscle rather than fat. Other experiments have suggested that CLAs can enhance immune function and reduce atherosclerosis, high blood pressure and inflammation. Quite a grab bag of positive findings! Pretty alluring, especially given that no significant side effects have been noted.

As one might expect, producers have been itching to add these compounds to regular foods so that they can then be promoted as having health properties beyond simple nutrition. While dietary supplements of CLAs made by chemically treating sunflower or safflower oils have been available for a couple of decades, their status as food additives has been in a regulatory limbo. Until now. With the FDA giving the go-ahead, producers are set to crank out cookies, eggs, yogourt, milk—and who knows what else—enriched with these "good" trans fats. Whether the promise of CLA enrichment is fulfilled remains to be seen. While the right dose may offer certain benefits, the Holy Grail has not been found. And if you want some CLA enrichment before fortified foods come our way, well, yak cheese and kangaroo meat are the way to go. Or you can take a supplement. Michael Pariza does—three grams a day. As for me, I'd like more evidence. But I just might dip my broccoli in Cheez Whiz. Geez . . . I can't believe I said that.

SOMETHING FISHY

IT'S A STRANGE WORLD. Health food stores promote dietary supplements of astaxanthin, claiming more powerful antioxidant benefits than with either vitamin E or beta carotene. It helps protect against the damaging effects of pollution, ultraviolet light and immune stress, they say. But in California, the state's supreme court ruled that private citizens can sue stores if they sell fish without declaring that astaxanthin has been added to their feed. What's going on?

Astaxanthin is a naturally occurring pigment found in a variety of algae that serve as food for krill, shrimp and crayfish, imparting an orange-yellow colour to these creatures. Algae produce astaxanthin for the same reason that many plant and animal species produce such antioxidants. They mop up the potentially harmful free radicals that are the by-products of metabolism and also offer protection from the damage that can be caused by ultraviolet light. This protection is transferred to the algae's predators, and then to the predators of those predators. Which explains why wild salmon develop their classic orange colour.

These days, however, most salmon are raised on fish farms. No, they're not genetically modified mutants that graze on the prairie, they are fish raised in penned-off areas of the ocean, where, instead of having to search for fresh krill, they can just lounge around waiting to be served a commercially concocted feast. But these pellets, made from fish too small and bony to be used for human consumption, lack the astaxanthin that gives wild salmon its colour. As a result, the flesh of farmed salmon turns out to be an unappetizing grey. And more grey means less green in the cash register. The answer to this little problem is to add astaxanthin to the feed.

The commercial production of astaxanthin is a huge industry, relying on three distinct processes. Sugar fermented by certain yeasts can produce the compound. It can also be extracted from specially grown algae. But the most economical, and therefore the most common, process relies on a fourteen-step chemical synthesis from raw materials sourced from petroleum. Actually, the name astaxanthin refers to any one of three very closely related compounds with very subtle differences in molecular structure. The ratio of the three produced by any of the commercial processes is the same, but differs from the ratio found in wild salmon. A technique known as high-performance liquid chromatography (HPLC) can be used to separate and quantify the three stereoisomers of astaxanthin, and hence determine whether the sample came from wild or farmed salmon.

Why should anyone care about this? Because wild salmon is prized more by consumers than farmed salmon! Some claim that they prefer the taste, but most who care about the origin of their salmon are concerned about their health. They've heard of studies showing that farmed salmon are higher in toxins such as PCBs than the wild variety. This may well be the case, since the meal fed to farmed salmon is made from fish often caught in more polluted parts of the ocean than where wild salmon feed. Whether these trace amounts of PCBs are of any health significance is debatable. My guess is that they are not. But what is beyond debate is that wild salmon fetch a higher price. So selling farmed salmon as wild is a lucrative, but obviously unethical, proposition. When it comes to a contest between ethics and profits, however, profits often triumph. So we shouldn't be too surprised that an investigation by *The New York Times* revealed that a number of stores were selling farmed salmon masquerading as wild. Samples were purchased from eight popular establishments and sent to a laboratory specializing in HPLC analysis. Six of the stores were found to be selling "fake" wild salmon.

Nobody likes to be duped in this fashion, especially not Californians, who tend to react to any type of perceived artificial meddling in their food supply with religious fervour. In many cases, the "meddling" was misunderstood, and consumers declared their outrage about their salmon being recklessly coloured with some artificial "chemical." I wouldn't be surprised if some of the anti-colourant demonstrators were popping astaxanthin antioxidant pills purchased in their local houses of worship, the health food stores. There is, of course, no health issue with the astaxanthin colourant in salmon, no matter what its origin. But people do have a right to know what they are buying. However, suing over whether a fish gets its astaxanthin colouring from dining on krill or from eating fish meal seems to be an overzealous and inappropriate use of the legal system.

If going to court over a colourant is deemed justified, just imagine the scenario when the litigious American consumer discovers that the farmed-fish industry is contemplating replacing some of the fish meal with a product made from grains and plant oils. As the population of the small fish used to make salmon feed declines, this is becoming a financially attractive proposition. And this substitution really could have a health consequence.

Norwegian researchers enlisted sixty coronary heart disease patients to study the health effects of eating salmon raised either on fish oil or on canola oil. For six weeks the subjects consumed three salmon meals a week and had their blood sampled for inflammatory substances as well as for various markers of cardiovascular disease. As was to be expected, the subjects who ate the fish reared on fish oils had higher levels of the heart-healthy omega-3 fats in their blood, but they were also found to have lower levels of two key markers of inflammation, vascular cell adhesion molecule-1 and interleukin-6.

If Californians think that the lack of labelling indicating the source of astaxanthin in salmon flesh is worth pursuing in court, they will certainly be most upset when they find out that their farmed salmon is being fed more like a cow than like a fish. But there is nothing illegal about this, so they may just have to drown their sorrows in a glass of carrot juice. Organic, of course. Perhaps fortified with a shot of astaxanthin.

DEMON DRINK

IT'S ANNOYING when facts get in the way of a good story. Like the one about Vincent Van Gogh mutilating his ear in a thujone-induced fit. Thujone is a naturally occurring compound found in wormwood, one of the plants used to flavour absinthe, the legendary liquor that enthralled artists and writers during the latter years of the nineteenth century and the early part of the twentieth. The "green fairy," as it was called, contained up to 75 per cent alcohol by volume, but its real kick was supposedly the "special" sensations induced by thujone. I have often told the story, pieced together from accounts in the scientific and popular literature, of Van Gogh being driven round the bend by thujone, with the compound possibly even contributing to his suicide. As a finale, I would recount how Dr. Paul Gachet, Van Gogh's personal physician, looked after his funeral and unknowingly adorned the grave with a wormwood bush that grew roots, eventually enveloping the casket. Van Gogh, I would say, was in the clutches of thujone in death, as he had been in life. Alas, recently uncovered facts suggest that we have to look elsewhere in an attempt to rationalize the artist's irrational behaviour. Thujone was not the culprit.

As it turns out, absinthe's naughty reputation isn't scientifically justified. Its status as a wicked, mischief-causing beverage can be traced not to carefully controlled experiments, but to a guinea pig, a murderer, some puritan prohibitionists and the French wine industry. Here's that story. Factual, at least as far as I can make out.

Absinthe was first formulated in Switzerland around 1790 by distilling an alcoholic brew infused with botanicals and herbs that included anise, hyssop, lemon balm, Florence fennel and *Artemisia absinthium*, or wormwood. The classic green colour is the result of adding chlorophyll extracted from herbs to the distillate. While it isn't clear who first came up with this concoction, we do know that the recipe ended up in the hands of Major Daniel-Henri

Dubied, who claimed that it enormously enhanced his sexual performance. The major then sold the recipe to his son-in-law, Louis Pernod, who seconded the bedroom effect, added a claim of "indigestion remedy," and began mass production in 1797.

Whether because of these purported special properties or its high alcohol content, absinthe became very popular, especially among artists, who took to indulging daily, and excessively, during *l'heure verte*. There was talk of enhanced creativity, but there were also murmurings about psychotic episodes, hallucinations and, according to some, permanent brain damage. Tracking down the origin of these accusations is difficult, but absinthe likely served as a convenient scapegoat for the inebriated antics associated with the bohemian community.

The flames of innuendo were fanned in 1864, when Dr. Valentin Magnan carried out the first-ever investigation into wormwood oil. Magnan placed a guinea pig in a glass cage with a sample of wormwood oil, and another one in a cage with a supply of alcohol. As might be expected, the latter animal lapped at the alcohol until he became drunk, but unexpectedly, the guinea pig exposed to just the vapours of wormwood oil went into convulsions. This primitive experiment drove the first nail into the coffin that would bury absinthe fifty years later. Magnan went on to isolate thujone from wormwood and confirmed its toxic potential by showing it caused convulsions followed by death in a dog. When the good doctor claimed to have evidence (never confirmed) that alcoholics who consumed absinthe were more likely to exhibit delusional behaviour and convulsions, the writing for the demise of absinthe was on the wall. The French wine industry, noting the increasing popularity of a competitor, was happy to jump on the anti-absinthe bandwagon alongside prohibitionists. And then came a catalytic moment.

A horrendous crime shook Europe in 1905. Jean Lanfray, a Swiss labourer, murdered his pregnant wife and two children in a drunken rage after she refused to polish his shoes. He had consumed seven glasses of wine, six of cognac, two crème de menthes, coffee with brandy and two shots of absinthe. Ignoring the stunning alcohol consumption, prosecutors claimed a clear case of "absinthe madness," a condition never scientifically demonstrated. Lanfray escaped the death penalty because "absinthe made him do it," but couldn't escape his own conscience—he eventually hung himself in his prison cell. The case set off an epidemic of moral indignation and triggered petitions to ban absinthe. By 1915 most countries, with the notable exceptions of the U.K., Spain, Portugal and Sweden, had made the sale of the drink illegal. The stated reason was that thujone in absinthe incited peculiar behaviour.

The myth that the original version of absinthe contained dangerous amounts of thujone persisted for almost a century—despite the fact that nobody had actually measured the thujone content of absinthe before blaming the drink for misdeeds. The amount thought to be present was actually estimated from what was known to be present in wormwood—and poorly estimated, as it turns out. In 2009 vintage bottles of absinthe were chemically analyzed for the first time and found to contain about 25 milligrams of thujone per litre, the same as in the currently available "low-thujone" versions, and far less than the 250–350 mg/L that had been previously claimed. We now know that even these levels are way below those that cause convulsions or hallucinations. Why, then, did so many absinthe imbibers experience such dreadful effects? Simple: they were plain drunk. So it seems that while thujone did not drive Van Gogh into the grave, the facts about this compound have buried my story about the artist being in its clutches. And if anyone wants to try some "pre-ban" absinthe, it is available. A bottle will set you back about three hundred dollars. But don't worry. It will not trigger any ear-mangling.

O.J. VERSUS RED BULL

"PYRIDOXINE!" "GLUCURONOLACTONE!" The words spew out with something close to contempt from the speaker's mouth as he ambles through a lab filled with brightly coloured liquids where technicians are seen to be using these ingredients to formulate some sort of beverage. Not the sort of chemicals we want to be defiling our body with, he implies, as he walks over and picks up a glass of orange juice. "Ingredients: fresh air, rain and sunshine," he declares. "Healthy, pure and simple." And with these words of wisdom the television ad for Florida orange juice comes to an end.

Before going any further, let me state that I think orange juice is a great beverage, and just the thing to start the day with. But that "pure and simple" beverage is composed of hundreds of different compounds, including some with tongue-twisting names like beta-cryptoxanthin, hesperitin-7-rhamnoglucoside and L-3-ketothreohexuronic acid lactone. Would it be comforting to learn that the last one is just the chemical term for vitamin C? And that the first one is a carotenoid, and the second a polyphenol, both of which are antioxidants linked to good health? The point, of course, is that the benefits or risks of a food or beverage are determined by the specific properties of its components, not by the number of syllables in their names. True, orange juice does not contain glucuronolactone, but that is not why it is a healthier beverage than Red Bull.

Why bring up Red Bull? Because that's the drink that features glucuronolactone as an ingredient and that apparently has cut into sales of orange juice. Glucuronolactone is an ominous-sounding synthetic compound, so it was a natural target for the orange juice TV campaign, especially given the ridiculous email that circulates about it. This much-forwarded diatribe maintains that glucuronolactone was an artificial stimulant developed in the 1960s by the

American government. Nonsense. Glucuronolactone can be found in the body as a natural product of glucose metabolism, and the amount found in a can of Red Bull is unlikely to have any negative effect. It is added to the drink with the insinuation that it increases energy, but that claim is unproven.

Red Bull is a curious beverage introduced to the Western world by Dietrich Mateschitz, an Austrian entrepreneur who encountered an "energy" tonic in Thailand called Red Water Buffalo. He thought that "bull" would sell better in the West and changed the name appropriately. Right on! Red Bull, inexplicably, now sells about two billion cans a year. The major ingredients were then, and are now, glucuronolactone, taurine (an amino acid), vitamins, sugar and caffeine. There is little evidence that Red Bull, which to me tastes like carbonated cough medicine, has any stimulant effect other than what can be ascribed to the caffeine it contains. How much caffeine? Roughly equal to that found in a cup of coffee and twice the amount in a cola beverage. Compared with orange juice, "energy drinks" like Red Bull are nutritional paupers.

Sure, orange juice is high in sugar, which is why low-carbohydrate diets like Atkins (happily on the wane) urge limited consumption. But if you cut out orange juice, you are also cutting out an excellent source of folate, potassium, flavonoids and carotenoids, all of which have been linked with health benefits. High blood pressure is more likely in people who consume less potassium. Indeed, a study at the famed Cleveland Clinic showed that drinking two glasses of orange juice a day resulted in a modest reduction in blood pressure. Researchers at the University of Western Ontario found that three glasses a day may help keep the doctor away. When they studied twenty-five patients with high levels of LDL, or "bad" cholesterol, they found that drinking orange juice increased HDL, the "good" cholesterol, by some 20 per cent, and that the ratio of LDL to HDL, a good measure of cardiovascular risk, decreased by 16 per cent.

And now we have a study that suggests that orange juice, in doses as little as one glass a day, may even stave off arthritis. In this case, the beneficial compounds appear to be carotenoids, the orange-coloured pigments found in a variety of fruits and vegetables which may reduce inflammation through their antioxidant effect. Researchers at the University of Manchester analyzed data from some twenty-five thousand subjects who had filled out dietary questionnaires. They compared people who eventually developed arthritis with those who did not and found that the average daily intakes of two specific carotenoids—namely beta-cryptoxanthin and zeaxanthine—were lower in subjects who came down with arthritis.

Of course, excess of anything is not good, either. Just ask the lady who ended up in hospital with an extremely high potassium level in her blood. Doctors could not figure out what was going on until she sheepishly admitted that she had been drinking about five litres of orange juice a day. Why? Because she had read about the silly Orange Juice Diet, which claimed that large doses can cleanse and rejuvenate the body. Well, instead of rejuvenating her, the huge dose of potassium almost killed her.

One last thing. A study at the University of Reading examined the effects of different kinds of breakfast on the IQ of children. Guess what: drinking orange juice in the morning improved their IQ. So maybe if people drank their OJ in the morning, they would be less likely to fall into the trap of a ridiculous diet craze and more likely to realize that there are better reasons to limit drinks like Red Bull than the fact that they contain glucuronolactone and pyridoxine.

By the way, pyridoxine is just vitamin B_6. Where can one get it besides Red Bull? Well, it's a natural component of orange juice!

DOES ORGANIC MEAN BETTER?

THE BATTLE HAS BEEN RAGING back and forth ever since synthetic pesticides and fertilizers were introduced into agriculture: is organic produce safer and more nutritious than the conventional variety?

Organic, of course, was once the norm. Until the twentieth century, all farming was "organic." If you wanted to fertilize your fields, you used manure or decomposing plant material. If you wanted to control insects, you used toxic, but of course "natural," compounds of arsenic, mercury or lead. Nicotine sulphate extracted from tobacco leaves killed insects effectively, and by the nineteenth century, pyrethrum from chrysanthemums was also available for insect control. Dusting crops with elemental sulphur was an age-old practice for reducing infestation by pests and fungi. And then, in the twentieth century, synthetic pesticides and fertilizers entered the picture. Why? Necessity, as has often been said, is the mother of invention. Crop losses were too great to feed the growing population, soils were being depleted of nutrients, and the toxic effects of arsenic, mercury and lead-based insecticides had become apparent.

Chemists rose to the challenge and developed fertilizers to replenish the soil and an array of pesticides to ward off insects and fungi. Yields increased, and the hungry were fed. At least in the Western world. With produce abundant and tummies full, we now had the luxury of turning towards other food-related concerns. Like the risks of the newfangled agrochemicals. After all, insecticides were designed to kill insects, so they obviously had toxic potential. Their effect on non-target species, such as interference with the egg-laying abilities of birds, began to raise questions about their effect on human health. Consumers began to hark back to the good old days when produce had been "chemical-free." They wanted uncontaminated, pesticide-free food grown without synthetic fertilizers. They wanted to go organic.

Some farmers complied. If that's what people wanted, they would go back to growing food the old-fashioned way. No pesticides, no synthetic fertilizers and none of those novel bogeymen, genetically modified crops. Sure, yields would be reduced, and the produce might look less appealing, but as long as consumers were willing to pay a premium, farmers would meet their needs. Indeed, consumers fearful of pesticide exposure were willing to pay more for organic produce, which they surmised would also be more nutritious. After all, doesn't Mother Nature know best?

A number of field trials were organized to compare the nutrient composition of organically and conventionally grown crops and produce. These focused mainly on antioxidant content, based on the general belief that it is these substances that account for the benefits of a diet high in fruits and vegetables. This is actually not as well established as most people think. While there is overwhelming evidence that a diet high in fruits and vegetables is healthy, there is no hard evidence that this is due specifically to antioxidant content. In theory, the assumption is reasonable, because antioxidants, at least in the laboratory, can neutralize free radicals, which have been linked with a variety of health problems. But fruits and vegetables contain hundreds of different compounds, and it isn't clear which ones are responsible for the health benefits. Studies with isolated antioxidants have proven to be disappointing.

Some, but certainly not all, studies have shown that organically grown foods are higher in antioxidants. This isn't surprising, because crops left to fend for themselves without outside chemical help will produce a variety of natural pesticides, some of which just happen to have antioxidant properties. And how much of a difference in antioxidant content is there between organically and conventionally grown foods? According to a four-year study carried out at the University of Newcastle, organic food is some 40 per cent richer in antioxidants. The researchers even suggest that this means we can eat fewer fruits and vegetables in our quest for good health, as long as they are organic. This is not a very compelling argument. Foods are extremely complex chemically, and measuring the amounts of a few antioxidants may not be a proper reflection of nutritional value. For that, we need feeding studies. Do rodents thrive on organic diets? Nobody knows. And are humans who eat organically healthier? Nobody knows.

There are some other questions that come to mind as well. What about disease-causing organisms that may be present in manure used as organic fertilizer? Or fungal metabolites, which are more likely to be found in organic foods because they are not protected by insecticides? Fumonisins, for example, produced by Fusarium moulds, are carcinogenic and have also been linked with birth defects in humans. Moulds take root where insects have damaged the crop. Such damage is less likely if the crops are protected through genetic modification. Inserting a bacterial gene that codes for the production of a toxin that has no effect on humans can protect these crops from insects. But, of course, genetic modification is not allowed in organic agriculture. Too bad, because if we look to increase nutrient content, this is the way to go. A line of genetically modified tomatoes, with almost eighty times more antioxidants than the conventional variety, has already been developed at the University of Exeter. Now, that is a far greater nutritional difference than between organic and conventional produce. Imagine the benefits we could have if organic farmers embraced genetic modification!

What, then, is the bottom line here? If cost is not an issue, organic may indeed be an appropriate choice. There is no doubt that it is a more environmentally sound practice. But for most people, cost matters, and if they commit to going organic all the way, expense and lack of availability may find them consuming fewer fruits and vegetables. Emphasis really should be on consuming at least seven servings of fruits and vegetables a day, not on whether these are organic or not. There is one more point to be made:

pretty soon, there will be ten billion people coming to dinner. And there is no way they are going to be fed organically.

JUST THIS ONCE

I SUSPECT IT WASN'T TOO DIFFICULT to get ethics committee approval for the study. Neither was it hard to find volunteers willing to wash down a slice of carrot cake with a milkshake. And basically, that's all the fourteen subjects had to do. Twice. The first time, the cake and milkshake were prepared with a saturated fat, and a month later the same meal was prepared with a polyunsaturated fat. Well, in truth, there was a little more to the task. The subjects had to donate blood samples before and after the experiment and also had to undergo a simple test to measure blood flow in their arms. This involved using a pressure cuff to stop flow and then measuring how quickly the blood vessels dilated to restore circulation once the pressure was released. The more readily a blood vessel dilates, the better shape it is in, and the common assumption is that blood vessels in the arm reflect the condition of vessels elsewhere in the body, including the heart.

Why the interest in carrot cake and milkshakes? Because these two delights can be prepared either with a saturated fat such as coconut oil, or a polyunsaturated fat like safflower oil. And it was the post-meal effect of these different types of fat that interested researchers at the Heart Research Institute in Sydney, Australia. Dr. David Celermajer, one of the principal investigators, had carried out previous work on HDL cholesterol, the substance that the lay press has justifiably labelled as "good" cholesterol. He found that there was more to the activity of HDL than just the commonly accepted mechanism of acting like a garbage truck that picks up excess cholesterol and prevents it from depositing in coronary arteries. HDL, Celermajer had found, was also capable of interfering with the production of "adhesion molecules" by cells in the inner lining (endothelium) of arteries. These molecules enable cholesterol to form deposits called plaque, which in turn cause arteries to be narrowed, restricting blood flow. A plaque that bursts can trigger the formation of a blood clot, which can then completely choke off blood flow and cause a heart attack.

Celermajer had a theory that the anti-adhesion activity of HDL was diet dependent and wondered if the composition of a single meal could have an effect. He chose to study this possibility by isolating HDL from the subjects' blood before the experimental meal as well as three and six hours after. He incubated endothelial cells in the lab in the presence of this HDL, then added a chemical (tumour necrosis factor alpha) known to stimulate the production of adhesion molecules and measured the ability of the various HDL samples to interfere with the process.

The results were surprising. While there was no significant difference in the amount of HDL present, samples taken from the volunteers after the unsaturated-fat meal had a much greater ability to prevent the formation of the nasty adhesion molecules than samples taken after the saturated-fat meal. Clearly, it wasn't the amount, but the quality of HDL that mattered! The tacit assumption has been that the higher the level of HDL in our blood, the greater protection it affords against heart disease. Now, thanks to this study, we are learning that there is more subtlety here, and the same level of HDL may be more protective or less protective, depending on its specific chemistry, which in turn may be determined by something as simple as our most recent meal.

The blood-flow experiments that were part of this study also yielded interesting results. After the saturated-fat meal, arteries in the arm took significantly longer to dilate after applied pressure had been released. This suggests that a meal high in saturated fat can impair the ability of arteries to dilate in response to a need for increased blood flow. And that is not a good thing.

The notion that a single meal can have such negative effects is certainly not welcome, as it seems to undermine the "just this once" argument. How often have we used that one before digging into a piece of cake? "I don't really feel like oatmeal with flax this morning. I'll have some sausages, eggs and hash browns. Maybe with a Danish. Just once can't hurt, can it?" Well, maybe it can. The saturated-fat content of the meal used in this study was roughly equivalent to that found in a cheeseburger, a large order of fries and a shake. Not an unusual combo for many. On the other hand, let's remember that the anti-adhesion effect of HDL was measured in cells in the lab, not in the volunteers' bodies. And as some critics have pointed out, the safflower-oil meal had far more vitamin E than the coconut-oil meal, and this, rather than the type of fat, may be responsible for the observed effects. But I would bet on the fat effect.

So, I think, would researchers at the University of Calgary, who had thirty healthy university students eat either a typical McDonald's high-fat breakfast or a low-fat, high-carbohydrate cereal breakfast, both with the same calorie content. After the meals, the students were subjected to various stressful tasks such as keeping their arms in ice water or doing math problems. Blood pressure, heart rate and the ability of blood vessels to dilate were all more favourable after the low-fat breakfast. And remember, these were healthy people!

But let me leave you with a bit of good news. It seems the bad effect of a high-fat meal usually peaks about four hours after the meal. And a study at Indiana University has shown that a forty-five-minute walk on a treadmill two hours after eating can prevent the effect. So if you're going to have that smoked meat sandwich with fries, take a good, long walk afterward to contemplate your sin.

YOU ARE WHAT MOMMY EATS

WE'RE GETTING FATTER. No doubt about it. The World Health Organization estimates that a seventh of the world's population is overweight, and about 300 million people can now be classified as obese. What's going on? The answer would appear to be pretty simple: we are eating more and exercising less. But that answer may be a tad too simple. Some researchers maintain that our increased calorie intake and decreased calorie expenditure are not enough to account for the "epidemic" of obesity we are witnessing. We had better have a look, they say, not only at the gluttonous amounts of processed foods we consume, but also at the packaging it comes in.

Until recently, such a notion would have seemed absurd. But now some intriguing research suggests that exposure in the womb to environmental chemicals, such as some of the fluorinated compounds used in grease-proof packaging, can be a predisposing factor for obesity. Of course, developing babies don't order pizzas, but their carriers have been known to make the odd call. And what mommy eats, the embryo eats. And if mommy eats hormone-like chemicals, baby may pay the price.

The usual suspects here include chemicals in detergents (nonylphenol ethoxylates), pesticides (atrazine, DDT, lindane), flame retardants (polybrominated diphenyl ethers) anti-fouling paints (tributyltin) and compounds leaching from plastics (bisphenol A, phthalates). Connecting these to weight gain seems off the wall, but it may not be. After all, hormones are commonly used to make cattle gain weight more quickly. While there is scant evidence that humans are being affected by tiny amounts of these endocrine disruptors, some animal studies do point to the possibility.

Retha Newbold's work at the U.S. National Institutes of Health is a case in point. Her interest was in diethylstilbestrol (DES), the classic estrogen-like compound that was once used to prevent miscarriage in women, unfortunately with tragic consequences. Their "DES daughters," as they came to be called, had an increased risk of clear-cell adenocarcinoma, a rare cancer. Newbold was investigating how DES might interfere with hormone systems when she made a surprising discovery. Injecting pregnant mice with tiny amounts of DES resulted in their offspring exhibiting unusual weight gain. Although food consumption and activity levels in the exposed mice were no different than in controls, by sixteen weeks of age, they had 25 per cent more body fat!

Diethylstilbestrol is not the only compound that has shown such an effect. Other researchers have connected in utero exposure to bisphenol A, phthalates and perfluorooctanoic acid (PFOA) with weight gain in rodents. Bruce Blumberg of the University of California actually coined a new term for such substances, calling them obesogens. His original interest was in the hormone-like effects of tributyltin, a fungicide used in paints, especially those used to protect ships' hulls from barnacles. Blumberg was taken aback when female molluscs exposed to the chemical grew male sex organs. Better see what it does to mice, he thought. Well, when pregnant mice were treated with tributyltin, their offspring showed unusual weight gain.

But, as the common saying goes, humans are not giant rodents. And, obviously, we can't expose pregnant women to suspected endocrine disruptors. So human evidence for the obesogen theory is hard to come by. Surprisingly, some support comes from, of all people, smokers. Smoking generally is associated with weight loss, but when women puff away during pregnancy (as unbelievable as that is), their offspring are twice as likely to be obese by the time they reach school age. In animal models, prenatal exposure to nicotine produces a similar effect.

What about post-natal exposure? Richard Stahlhut and colleagues at the University of Rochester have linked higher levels of phthalates in men's urine with more belly fat. Previously, low levels of testosterone have been associated with abdominal obesity, and phthalates, at least in animal studies, depress testosterone. So a connection between phthalates and jiggling bellies is plausible. And the connection may not be limited to men. Recently, researchers at Mount Sinai School of Medicine in New York investigated exposure to phthalates by looking for the metabolites of these compounds in the urine of some four hundred girls in East Harlem. The heaviest girls had the highest level of phthalate metabolites in the urine. Of course, such an association does not prove that phthalates are responsible for weight gain. Maybe the heavier girls ate more high-calorie processed foods, which would expose them to more phthalates used in packaging.

Obesity is a complex phenomenon, with multiple potential contributing factors. For example, studies have shown that sleep deprivation can lead to weight gain, and data show that the average daily sleep has decreased over the past few decades from nine hours to seven. Smoking reduces weight, and there are fewer people smoking. Some drugs, especially antidepressants and antipsychotics, produce weight gain, as do antidiabetics and beta-blockers. The age at which women have babies is also a factor. Having an older mother is a risk factor for obesity, and since 1970 the average age of first pregnancy has increased by about two years.

Believe it or not, even air conditioning and heating may be connected to obesity. Maintenance of body temperature requires energy expenditure, meaning that more calories are burned when we have to cope with high or low temperatures. Interestingly, in the southern U.S., which has an extremely high obesity rate, the percentage of homes with central air conditioning has increased from about 30 to 75 per cent since 1980. When people are comfortable, they eat more.

Taking all this into account, jumping on the endocrine-disruptor/weight-gain bandwagon is premature. It is more important to worry about the quantity of food in the package we put into our mouth than about the quantity of chemicals the package puts into the food. And until the laws of thermodynamics are repealed, the way to lose weight is still the old-fashioned way: eat less and exercise more.

HEALTH
ISSUES

LISTERIA HAPPENS

I DONNED A WHITE LAB COAT, put on a hair net and scrubbed my hands thoroughly. I wasn't sure what to expect once the doors swung open, but I think "cleanliness" best encapsulates my first impression. Gleaming stainless-steel equipment, employees clad in white smocks, sinks with foot pedals for hand-washing and technicians swabbing for microbial contamination. A hospital operating room? No—a plant for processing meat into a variety of cold cuts and sausages.

In light of the listeria issue that terrified Canadians in 2008, I wanted to get a first-hand look at just how a modern food-processing plant operates. So off I went to visit a Montreal company that specializes in Italian-style deli meats. I was basically interested in how a large commercial establishment addresses that bane of food production, contamination with disease-causing bacteria. I came away with a good taste in my mouth, both figuratively and literally.

First, a word about bacteria. These single-cell microbes were the first form of life, appearing on Earth roughly four billion years ago.

They are everywhere. They're in our soil, in our water, in our air, in our food—and in us. In fact, if you want to talk numbers, we are more bacteria than human. Our body contains roughly a hundred trillion cells, but we host ten to twenty times as many bacteria. Indeed, we, as well as other animals, are a veritable hotel for more than five hundred different species of bacteria. They live on our skin, in our noses, our throats, but mostly in our intestines, where they have a ready supply of nutrients from the food we ingest. That's where most of the one or two kilograms of bacteria that we harbour set up shop. And they do pay rent: in return for a cozy, nutrient-rich environment where they can reproduce every twenty minutes, bacteria help digest our food. Some even produce vitamin K for us, as well as biotin, one of the B vitamins. And as they digest food, they crank out butyric acid and other short-chain fatty acids with anti-cancer properties. Given the massive number of bacteria that populate our guts, it should come as no surprise that these microbes make up roughly 60 per cent of the dry mass of feces. Understandably, then, exposure to fecal matter is a prime means of bacterial contamination.

About 85 per cent of the bacteria that live in our guts are beneficial, but the rest are pathological, meaning that they can cause disease. The names are familiar: salmonella, E. coli, staphylococcus, clostridia, campylobacter and, of course, listeria. Since all bacteria compete for the same food supply, these nasties are normally kept in check by the overwhelming number of beneficial bacteria. But an influx of pathogenic bacteria from an outside source such as food can overpower the helpful bacteria and wreak havoc. Diarrhea, vomiting and fever are the body's attempts to get rid of the invaders. These defence mechanisms generally yield favourable results in a few days, but on occasion, some of the bacteria may infiltrate the bloodstream from the gut and cause major problems such as blood poisoning or meningitis, signalled by a stiff neck and headache. In people with weakened immune systems, as well as in the very old or the very young, such infections can be lethal, especially in the case of infection with listeria. It is difficult to determine how many cases of food poisoning occur each year, since most people do not seek medical help for a few days of malaise, but a reasonable estimate is two million cases a year in Canada, with about thirty deaths. Given that we consume about forty-five billion meals a year, the chance that a single meal can be lethal due to bacterial contamination is about one in 1.5 billion.

Of course, even a single death is too many, especially if it is preventable. That's why so much effort is put into reducing the risk of bacterial contamination of our food supply. Meat is a haven for bacteria, especially if it has come into contact with fecal matter during slaughter, a very likely scenario. The invaders can be killed by appropriate heating, drying or salting, all common processes used to make deli meats. But these processes do not preclude subsequent contamination during the slicing, packaging or distribution stages. Over the last few decades, researchers have analyzed every step in the production chain and have evaluated the risk of contamination at each stage. A system known as Hazard Analysis Critical Control Points (HACCP) has been introduced and, if followed, it minimizes the risk of contamination. HACCP was developed by NASA for the U.S. space program—obviously, diarrhea in space is most unwelcome.

Unfortunately, no matter how clean a facility, no matter how closely HACCP is followed, contamination can occur. Given the astonishing extent of modern food production, it is surprising how infrequently such problems arise. That's why they make the news! What could have happened to cause, for example, the listeriosis outbreak at a Maple Leaf Foods plant in 2008? We may never know. What we do know is that Maple Leaf, like other producers, continuously swabs drains, working surfaces, equipment and product to test for the presence of microbes. One of the tests conducted at the Toronto plant revealed the presence of listeria monocytogenes and triggered a recall. Unfortunately, by the time the word was out, contaminated product had been consumed, with some tragic results for a dozen or so people who had weakened immune systems owing to age or disease. The risk to the general population was minuscule, certainly not warranting any panic.

Somehow—exactly how remains a mystery—bacteria from raw meat found their way into product after it had been heat-treated. Such an outbreak can begin when a single sample of finished, processed product becomes contaminated and is then sliced. Bacteria are capable of sticking to the stainless-steel parts of a slicer in the form of a biofilm and can contaminate hundreds of subsequent slices. These packaged slices are then delivered to stores, where the bacteria can multiply. Indeed, in the case of the Maple Leaf Foods plant, listeria bacteria were eventually detected in a slicing machine even though it had been maintained according to the manufacturer's instructions.

One of the problems with listeria is that the microbe, unlike others, readily reproduces at refrigerator temperatures. By the time an unsuspecting consumer eats a sandwich, the meat can be loaded with bacteria. Listeriosis, here we come!

Can anything be done to reduce the risk of such outbreaks? Sure. Post-packaging pasteurization, as is done in the U.S., reduces risk. So does holding product back from distribution until microbial testing has shown it to be clear. Bacteriophages, viruses that destroy bacteria, can be added to cold cuts, and irradiation can eliminate bacteria. Inspection methods can always be improved. The question comes down to how much effort is warranted to reduce a very small risk to an even smaller one. And let's remember that processors live in constant fear of an outbreak such as we witnessed with Maple Leaf Foods, because it can destroy a company. That's why they do everything in their power to keep their facilities as immaculate as possible. But humans are humans and nature is nature, and stuff happens. Sometimes, bad stuff.

I'm often asked how much I worry when something like the listeria outbreak comes along. Since I'm not pregnant, immune-compromised or old (although I suppose that is a matter of opinion), I do not fall into the high-risk group. In any case, I don't eat more than a hundred grams or so of deli meats a week, mostly because these are high in fat and salt and excessive consumption has been linked to bowel cancer. I do practise the usual recommended preventive procedures: I wash my hands often, wash cutting boards and counters with soap and hot water, wash fruits and vegetables thoroughly, clean the fridge frequently, refrigerate foods promptly, and trust in HACCP and the extensive testing for microbial contamination that food producers carry out. And I'm careful when I drive or walk to the store, since statistically the chance of my being injured in an accident is greater than the risk of being harmed by any food that I might purchase.

FEEDING THE BRAIN

MOST NORTH AMERICANS DON'T EAT RIGHT. Too much of the wrong kinds of fats, too many sweets and not enough fruits and vegetables. Maybe that's because we're just not smart enough to know what to eat. Maybe we need to feed our brains better! How does a meal of baked fish with nuts, a side of cooked lentils and a spinach salad sound? You can throw in some milk chocolate for dessert. Not only may this combo boost mental performance, it may even slow age-related mental decline.

Let's start with the fish. It seems that those old wives who plied their kids and husbands with fish to make them smarter were pretty smart themselves. Unwittingly, they may have been supplying their loved ones' brains with those now-famous omega-3 fats. These, as it turns out, are not only essential building blocks for the structure of brain cells, they are also needed for those cells to function normally. Ancient folklore has now been backed up by modern science. When researchers at Rush University Medical Center in Chicago interviewed some four thousand seniors every three years about their dietary habits and assessed their mental function, they found that people who ate fish at least once a week had a 13 per cent lower rate of cognitive decline than those who ate fish less frequently. What does this mean? One way of interpreting the result is that fish eaters are mentally three to four years younger than abstainers.

Dr. Nicolas Bazan of Louisiana State University thinks he can explain the protective effect of fish consumption. He's probably a fish eater. Bazan compared the brains of people who had died from Alzheimer's disease with those of people who had died from other causes and found that higher concentrations of docosahexaenoic acid (DHA), the main omega-3 fat found in fish, reduced the inflammation and toxicity caused by the buildup of beta amyloid protein in brain cells. This protein is widely regarded as the culprit in Alzheimer's disease. But Bazan discovered something else as well: DHA not only protects against the ravages of beta amyloid, it can actually allow brain cells to survive longer. It is the raw material that enzymes use to produce neuroprotectin DI, a compound responsible for turning on genes that promote the survival of brain cells.

Indeed, Bazan's group found that areas of the brains of Alzheimer's patients known to be critical for memory and cognition were particularly low in neuroprotectin DI. Furthermore, the researchers discovered that adding DHA to cell cultures designed to mimic the effects of aging spurred the production of the neuroprotective factor. All of this suggests that we should ensure our brain is well fuelled with the right kind of oil. But just how much DHA is adequate? Certainly not the average North American intake of 60 to 80 milligrams a day. We need more like the 200 to 300 milligrams found in a serving of fish. And if you want a really whopping amount of DHA, well, caviar is your answer. A 100-gram serving contains a stunning 4,000 milligrams of DHA! Of course, DHA supplements are also available and are cheaper than caviar.

Fish are not the only source of omega-3 fats. Flaxseeds and nuts are rich in alpha-linolenic acid (ALA), an omega-3 fat that also is beneficial, although probably less so than fish oil. Some ALA, however, is converted in the body to DHA, so nuts certainly are a welcome addition to a "brainy" diet. This is especially the case if we consider that nuts are an excellent source of selenium. When French researchers analyzed blood samples periodically taken from 703 subjects over nine years, they found that cognitive decline was associated with decreases of selenium in the blood over time. These results mesh with previous findings linking selenium-containing proteins to brain function. Best sources of selenium? Brazil nuts and tuna.

Now on to the spinach and lentils. Folic acid is the nutrient of interest here. Why? Because it is involved in the metabolism of homocysteine, a naturally occurring substance that, when found in abnormally high levels in the blood, has been linked with impaired brain performance. The more folic acid there is in the blood, the less homocysteine, goes the argument. And there may be something to this. Maria Corrada and Claudia Kawas of the University of California analyzed the diets of 579 non-demented men and women over the age of sixty, carefully noting all the foods they ate and dietary supplements they took. Eventually, a tenth of the participants developed Alzheimer's disease. What was the difference between those afflicted and those who remained healthy? Folic acid intake! Participants who consumed at least 400 micrograms a day were far less likely to become Alzheimer's victims. Spinach and lentils are great sources of folate, containing about 150 and 200 micrograms per serving.

The observation that increased intake of folic acid is linked with a reduced risk of Alzheimer's brings up the possibility of preventing the disease by upping our intake of folic acid. Since force-feeding volunteers with spinach and lentils on a daily basis is hardly a viable option, scientists looked towards folic acid in pill form. When Dutch researchers randomly assigned 818 older adults with elevated homocysteine levels to take either 800 milligrams of folic acid or a placebo daily for three years, they found improved memory and speedier information processing in the folic acid group.

While many people may struggle with dietary guidelines urging them to eat more fish and spinach, few would object to the prospect of boosting brain power with milk chocolate. Dr. Bryan Raudenbush of Wheeling Jesuit University gave student volunteers either 85 grams of milk chocolate, 85 grams of dark chocolate, 85 grams of carob or nothing before asking them to undergo various cognitive tests. Believe it or not, scores for visual and verbal memory were significantly higher for the milk chocolate group! I can't figure out why that should be so. Maybe if I ate some spinach, or some milk chocolate–covered nuts . . .

TOO MANY COCKTAILS

THE NUMBER OF CHEMICAL REACTIONS going on in our body at any moment is astounding. Amino acids are combining to form proteins, carbohydrates are generating energy, antioxidants are eliminating free radicals, DNA molecules are replicating themselves, antibodies are targeting intruders for destruction and a staggering array of enzymes are producing chemicals ranging from insulin to hemoglobin. When everything is going smoothly, we are healthy. But what happens when we start introducing substances that can gum up the complex machinery? That's where toxicology enters the picture. And what a complex picture it is.

Today, the number of foreign chemicals to which we are potentially exposed is truly astonishing. Pesticides, cleaning agents, drug residues, solvents, food additives, plasticizers, heavy metals and fire retardants are just a few examples of the chemicals that can find their way into our bodies. While we do know a fair bit about the potential toxicity of these substances individually, we know very little about what happens when they are present in some combination. That's why many environmental groups are clamouring for toxicological testing not only of individual compounds, but of various combinations. A noble notion, but unrealistic.

Nobody contests the possibility that the presence of one chemical may drastically alter the way we respond to another substance. Alcohol can increase the effect of barbiturates; fish-oil supplements can cause bleeding in someone taking the anticoagulant warfarin; zinc lozenges can interfere with the absorption of copper or calcium; and smoking increases the risk posed by exposure to asbestos. But systematically testing combinations of chemicals for toxicity, a seemingly logical thing to do in theory, is impossible in practice. There are just way too many such combinations.

This doesn't mean that we should not try to learn more about the "cocktail" effect. Almost all the information we have about the toxicity of pesticides, for example, comes from studying individual chemicals in animals. It is certainly possible that compounds that have no effect individually can cause a problem when combined. For years, researchers have suspected a link between pesticides and Parkinson's disease, mainly because of a higher incidence in agricultural areas. No single chemical culprit has been found, but that may be because a combination effect is involved. Researchers at the University of Rochester School of Medicine and Dentistry think this could well be the case. They injected mice with the herbicide paraquat and the fungicide maneb, both individually and in combination. After the experiment, the animals were sacrificed and their brain analyzed for signs of Parkinson's. While the individual compounds had no effect, the combination resulted in a reduced production of the brain chemical dopamine, a characteristic of Parkinson's disease. The researchers hypothesize that the offender is paraquat, which normally is not readily taken up by the brain. Its absorption, however, may be promoted by the presence of maneb. But is injection of these chemicals into mice an appropriate model for human exposure? Nobody knows.

And are chickens an appropriate model to study "Gulf War disease"? A large number of veterans from the first Gulf War (of 1990–91) have reported symptoms ranging from memory loss and unusual fatigue to shortness of breath and tremors. Scientists at Duke University wondered if the cause was some sort of chemical exposure. But to what? They considered three substances that were widely used to protect American soldiers during the war. The main concern was the possible use of nerve gas by Saddam Hussein, so many soldiers were administered the anti–nerve gas agent pyridostigmine bromide. Another fear was insect-borne disease, such as malaria; as a result, insecticides like DEET and permethrin were widely used. Each of these chemicals had been individually tested in animals and had been cleared for use in humans—in fact, at even higher doses than those actually administered to the soldiers. Possible combination effects, however, had not been studied, at least not until the Duke researchers tested the combination in chickens and found that, while the individual compounds showed no toxicity, various mixtures of two of them caused shortness of breath, stumbling and tremors. The most severe signs were seen when all three chemicals were combined. Again, the chicken study hardly proves that Gulf War Syndrome is the product of a chemical combination effect, but it is an interesting observation. So is the one made by biologist Tyrone Hayes of the University of California.

Hayes looked at the effect of some commonly used insecticides, herbicides and fungicides on tadpoles. Separately, the chemicals had no effect, but the combination, at levels found in many cornfields, made the tadpoles more susceptible to infection and impaired their development into frogs. Rick Relyea at the University of Pittsburgh found something even more interesting. The presence of the pesticide carbaryl in water doesn't bother bullfrog tadpoles, at least not until predatory newts are around—then there is a problem! The mortality rate among the tadpoles increases dramatically when a combination of the smell of newts and carbaryl are present at the same time, while tadpoles keep frolicking happily if exposed to either the smell or the pesticide individually.

Since we are not tadpoles, it is hard to know how relevant such a study is to humans. The same goes for one conducted at the University of Liverpool, in which a combination of the flavour enhancer MSG and the artificial colourant brilliant blue (also known as FD&C #1) stopped the growth of mouse nerve cells in a test tube. Neither food additive had a significant effect by itself. Let's remember that "natural" chemicals also enter this complex mixture. McGill researchers Robert Segal and Louise Pilote report the case of a woman being treated with the anticoagulant Coumadin who developed severe internal bleeding after drinking chamomile tea. Obviously, there are potentially dangerous chemical combinations out there, but we just cannot find them all. Science does not have all the answers.

BITTER SWEET

"YOU HAVE A TOUCH OF SUGAR." Not an unusual comment to hear from a physician. Especially if you are over fifty and starting to have trouble seeing your toes. Most people don't pay much attention. Why should they? They feel fine. And after all, a "touch" isn't diabetes. They may think about cutting back on eating sweets for a few days, and then they'll forget about that "touch of sugar." But their bodies won't. That little extra sugar is likely to become a lot of extra sugar, which heralds the presence of type 2 diabetes with all of its complications. And there is more bitter news about that "touch of sugar." Even if it doesn't evolve into diabetes, it still increases the risk of a heart attack or a stroke.

Chemically, the term "sugar" refers to a class of simple carbohydrates that have a sweet taste. Sucrose, lactose, fructose and mannose are all sugars. But that "touch of sugar" refers to one specific sugar: glucose. And it is an important one. Without it, you can't move a muscle. Without it, you can't think. But unfortunately, it can also kill you. Glucose, when properly absorbed into a cell, is burned as fuel to supply the body's energy needs. But if it runs wild in the bloodstream, it reacts with various proteins and fatty substances (lipids) to form advanced glycation end products (AGEs), nasty substances that can wreak havoc with the workings of the kidneys, the eyes, the nervous system and the coronary arteries.

Indeed, a high level of glucose in the bloodstream is a recipe for disaster. A disaster called diabetes. And it is reaching epidemic proportions—particularly the variety known as type 2, which makes up about 90 per cent of all cases. Predictions are that if we don't do something about it, we will soon be looking at more than thirty million North Americans afflicted with type 2 diabetes, most of whom will die from some complication of the disease. Luckily, there are ways to curb the carnage. In many cases, it comes down to a simple matter of losing weight. Unfortunately, that "simple" refers to the concept, not to the practice.

Ancient Chinese physicians already knew that if ants gathered around spilled urine, it was a bad sign for the urinator. Ants like sweets, and glucose fits the bill. The compound ends up in the urine when the kidneys do their best to remove excessive amounts from the blood. In fact, one of the first signs of diabetes is frequent urination (polyuria) as the kidneys struggle to eliminate glucose. With all that loss of urine, extreme thirst develops (polydipsia), another indicator of the disease. While these symptoms are no more than annoyances, serious problems may lie just around the corner. Blurry vision, fatigue, sores that don't heal, burning sensations and numbness in the extremities soon signal that glucose is engaging in reactions with essential biomolecules, impairing their performance.

Where does this double-edged sword, this glucose, come from? We eat it. Lots of it. Not directly, but as a component of a wide array of carbohydrates. Starches, table sugar (sucrose) and milk sugar (lactose) all yield glucose when broken down by enzymes in the small intestine. From here, glucose is absorbed into the bloodstream and, in turn, into cells. That is, if all goes well. In diabetes, it does not. The problem is not the influx of glucose into the blood from the intestine—if that were the issue, we would all be suffering from the disease. The problem lies in the entry of glucose into cells. In type I diabetes, which usually strikes early in life, the pancreas stops producing insulin, the hormone that cells (other than those that make up the nervous system) need to absorb glucose. In this case, the only recourse is the injection of insulin. Type 2 diabetes appears later in life, and is usually associated with being overweight. The pancreas still produces insulin, but cells cannot use it properly; they become insulin resistant, and glucose levels in the blood climb. As the pancreas struggles to churn out more and more insulin in an attempt to force glucose into cells, it eventually falters and full-blown diabetes sets in.

Why are some people affected, and others not? What causes the destruction of insulin-producing cells in the pancreas in type I and the failure of cells to respond to insulin in type 2 diabetes? Let's get one thing straight. Eating sweets or the wrong kind of foods does not cause diabetes. Poor dietary habits, however, can lead to obesity, which is a risk factor for type 2 diabetes. There are theories aplenty about the causes of diabetes, but they are just theories. Perhaps a virus or bacterium triggers an immune reaction that spins out of control and ends up attacking the pancreas. Perhaps some sort of toxin is involved; this theory is gathering steam with the increasing incidence of type 2 diabetes. A few studies have found an association between diabetes and exposure to chemicals such as PCBs, dioxins, methyl mercury or bisphenol A. These may have a direct effect on either insulin production or use, or perhaps they increase fat retention and lead to obesity, which in turn leads to diabetes.

While researchers have not been able to unravel the causes of diabetes, it is clear that heredity plays a part. If one of a set of identical twins develops type 1 diabetes, the chance that the other one also will is about 50 per cent. With type 2, the odds rise to 80 per cent. But clearly, environmental factors such as diet, infections or toxins play a role. While the causes of diabetes remain somewhat of a mystery, we have learned one thing for sure. The physical damage to the body is caused by excess glucose in the blood. This is not as obvious as it may seem. After all, wrinkles and osteoporosis go hand in hand, but wrinkles do not cause the condition. Similarly, there was a possibility that high levels of glucose were associated with diabetes, but that glucose was not responsible for the symptoms of the disease. In the early '90s the Diabetes Control and Complications Trial settled that question once and for all.

Some fifteen hundred patients with type 1 diabetes were divided into two groups. Half the volunteers took their regular insulin shots and monitored their blood glucose once a day. They made no adjustments as long as they felt well. The others tested their blood sugar at least four times a day, sometimes even in the middle of the night, and were given a specific range of numerical values to shoot for (below 6.7 millimoles per litre of glucose before meals and no more than 10 millimoles per litre). Diet and insulin doses were adjusted to meet this goal. Results were so remarkable

that the trial was stopped earlier than planned to allow participants to take advantage of the benefits of careful monitoring. In
the group that strictly controlled blood glucose, vision problems
were reduced by 76 per cent, kidney disease by 50 per cent and
neurological complications by 60 per cent. Obviously, monitoring sugar levels is critical in diabetes, but what can you do to keep
the levels under control? And what can you do to prevent type 2
diabetes from developing in the first place?

Let's get down to some numbers. There's only one way to
know the status of sugar (glucose) in your blood: measure it. If
after an overnight fast the concentration of glucose is more than
7.0 mmol/L, you've got diabetes. If it is between 6.0 and 6.9
(5.5 and 6.9 in the U.S.), you're dealing with prediabetes, sometimes referred to as glucose intolerance or insulin resistance.
Another way to measure blood glucose is by means of a glucose
tolerance test, which involves testing two hours after drinking a
glucose-laced solution. Above 11.0 mmol/L indicates diabetes,
between 7.8 and 11.0 is prediabetes. These tests provide an important snapshot of blood glucose, but a different kind of test is
needed to determine average glucose levels over a period of time.

The hemoglobin A1c test is based on the reaction of glucose
with proteins that are part of hemoglobin, the oxygen-carrying
molecule in red blood cells. The more glucose in the blood, the
greater the chance that these proteins are glycated. Since red blood
cells have a lifetime of three to four months, determining the percentage of hemoglobin that is glycated gives a picture of glucose
control over several months. More than 6 per cent glycation signals
a problem.

That "touch of sugar" the doctor talks about is prediabetes.
And it isn't rare. We're talking about fifty million people in the
U.S. and Canada, most of whom will go on to develop full-blown
diabetes unless they take action. And thanks to a landmark
study published in the *New England Journal of Medicine* in 2002, it

is clear what that action should be. The Diabetes Prevention Program (DPP) enlisted more than three thousand prediabetics with an average age of fifty-one. Almost all were overweight and sedentary, which wasn't surprising since elevated blood sugar has long been associated with these characteristics. A third of the subjects were treated with metformin (known by the brand name Glucophage), a medication that lowers blood sugar, a third with a placebo and a third were given a lifestyle-modification program. This last group exercised two and a half hours a week and was instructed in ways to reduce body weight by cutting the fat content of their diet.

The results were so dramatic that the study was halted a year earlier than planned. After just three years, 30 per cent of the subjects on placebo had developed diabetes—compared with 22 per cent of those on metformin, but only 14 per cent of those who had modified their lifestyle! And the average weight loss of the third group was only about four kilograms! The conclusion is obvious: type 2 diabetes is preventable, as are the risks that go along with it. High blood sugar damages the eyes, the nervous system, the kidneys and the cardiovascular system.

But now for the truly scary part. The risk of heart disease and stroke kicks in well below the blood levels of glucose that define diabetes. That surprising finding came from a fascinating study by researchers at the Harvard School of Public Health who gathered data on blood sugar from fifty-two countries, along with information about the incidence of heart disease and stroke. Sophisticated number-crunching revealed that blood glucose in the prediabetic range was a leading cause of cardiovascular death. Higher-than-optimal blood sugar accounts for more than three million deaths a year.

Obviously, controlling blood sugar is very important, even for non-diabetics. Especially so for anyone diagnosed with metabolic syndrome, a collection of disorders that often culminates in diabetes and heart disease. Metabolic syndrome encompasses high blood pressure, high triglycerides, low levels of "good" cholesterol (HDL), elevated blood sugar, and a waist size greater than 40 inches for men and 35 inches for women.

All right then, we have to control our blood sugar, but how do we do it? Intuitively, we immediately think of reducing our sugar intake. Not a bad idea, but too simplistic. Other foods can actually drive up blood glucose more effectively than sugar. What really matters is the total carbohydrate content of a food, and the rate at which it releases glucose into the bloodstream. The glycemic index measures the degree to which blood sugar is elevated in response to eating an amount of a food that contains 50 grams of carbohydrate. But alone, this is not very meaningful. Watermelon has a high glycemic index, but you have to eat a lot of melon to consume 50 grams of carbohydrate. Potato chips have a lower glycemic index, but you can easily consume 50 grams of carbs. That's why the best measure in terms of controlling blood glucose is the glycemic *load*, which takes into account the glycemic index *and* the available carbohydrates.

Which, then, are the foods with high glycemic loads that we should curtail? Potato chips, for sure. Unfortunately, other types of potatoes too. Candy bars. Anything made with white flour. These flood the blood with glucose, which then reacts with various proteins to form those advanced glycation end products (AGEs), which as we have seen cause the health problems. Recently we have also learned that grilling, broiling or frying foods causes similar reactions in the food itself, and that the AGEs can be absorbed into the bloodstream. The moral of the story is exercise, reduce fats, reduce cooking temperatures and reduce the consumption of anything "white" (low-fat dairy products excluded—these actually may reduce the risk of diabetes), and you'll reduce that "touch of sugar" in your blood. And you'll increase your chances for a sweet life.

OSTEOARTHRITIS SUFFERERS SPOILT FOR CHOICE

WHAT DO THE FOLLOWING HAVE IN COMMON: cat's claw, devil's claw, ginger, green tea, MSM (methylsulphonylmethane), green-lipped mussels, SAMe (S-adenosylmethionine), boswellia, *sangre de grado*, olive oil, hydrolyzed collagen, cider vinegar, tart cherry juice, gin-soaked raisins, fish oil, vitamins B, C, D or E, calcium, copper, stinging nettle, Certo, Willard Water, hot pepper extract, magnets, aloe vera, acupuncture, vegan diet, WD-40, cetyl myristoleate, sea cucumber, sesame seeds in the navel, bathing in manure, standing naked under the full moon, chondroitin sulphate and glucosamine? All right, so the last two give it away. They are all non-prescription remedies that supposedly treat osteoarthritis. What else do they have in common? None of them is backed by compelling scientific evidence. But some do offer hope.

Any chronic condition that significantly affects more than 10 per cent of the population, and for which conventional medicine offers no solution, is destined to generate a host of alternative treatments. Alternative, of course, does not necessarily mean ineffective. It just means that insufficient evidence is available to incorporate the regimen into mainstream medicine. All of the treatments above have their champions, and all come with testimonials galore. But unfortunately, the plural of anecdote is not data.

Osteo means bone, *arth* means joint and *itis* means inflammation. So osteoarthritis literally means inflammation of the joint. It is somewhat of a misnomer because the joint is not necessarily inflamed. By definition, inflammation encompasses swelling, warmth, redness and pain. Osteoarthritis is painful, to be sure, but it can occur without the other symptoms. Or with them. That's because osteoarthritis is actually a complex condition, not adequately described by the usual simple definition of "breakdown of cartilage."

Cartilage is the protective layer of tissue that covers the ends of bones, helping to absorb stress when a mechanical load (such as excess weight) is applied and providing a smooth surface for low-friction movement of the joint. It is composed of a matrix of collagen (a protein) and molecules called proteoglycans, in which various carbohydrates are linked to a protein backbone. Embedded in this matrix are chondrocytes, the cells that produce the matrix components. Like other living cells, chondrocytes undergo continuous turnover, meaning they are constantly being generated and destroyed. Matrix components also undergo turnover, being synthesized by chondrocytes and broken down by enzymes called metalloproteinases.

As long as chondrocyte and cartilage matrix turnover is balanced, joints remain healthy. But when there is a reduced formation or increased breakdown of cartilage matrix, the result is osteoarthritis, with its characteristic symptoms of joint pain, tenderness, limitation of movement and varying degrees of inflammation. The mechanical properties of the bone lying just under the cartilage can also affect the rate at which cartilage degrades. Bone of lower mineral density is less supportive, explaining why osteoporosis is often associated with osteoarthritis.

The factors involved in osteoarthritis suggest a variety of possible treatments. Weight loss as well as painkillers (acetaminophen is a common choice), non-steroidal anti-inflammatory drugs (NSAIDs) or cortisone injections can help control symptoms but of course do not address the underlying disease. That can only be addressed by somehow affecting the synthesis or breakdown of the cartilage matrix. The simplest idea that comes to mind is to provide chondrocytes with an increased supply of the raw materials they need to make proteoglycans, the critical components of cartilage matrix. This is where things get complicated, but bear with me.

Proteoglycans are composed of proteins and carbohydrates referred to as glycosaminoglycans (GAGs). Chondroitin sulphate is a typical GAG needed for cartilage formation, and glucosamine is a precursor for the formation of several glycosaminoglycans. Not surprisingly, then, both chondroitin sulphate, readily available from bovine cartilage, and glucosamine produced from crab shells have been extensively investigated as treatments for osteoarthritis.

On first glance, the use of these substances makes sense, but there is a theoretical glitch. Chondroitin sulphate is a complex macromolecule, and it is hard to imagine it surviving digestion and passing intact into the bloodstream. Glucosamine, on the other hand, is a simple compound, and ingestion can probably increase blood levels. But the compound is actually readily synthesized in the body from glucose, which of course is abundant in the diet. There is no reason to think that anyone has a deficiency in glucosamine needed to make cartilage. Collagen supplements have also been promoted as an aid in battling osteoarthritis, and they may increase the levels of amino acids in the blood that chondrocytes need to make the collagen part of cartilage matrix. But again, there is no shortage of supply of such amino acids in our diet. Of course, what matters is not theory but clinical evidence. And unfortunately, that evidence is rather weak. Although anecdotes abound, proper controlled clinical trials have shown that neither chondroitin nor glucosamine perform better than a placebo in slowing the rate of cartilage loss.

The Glucosamine/Chondroitin Arthritis Intervention Trial followed hundreds of subjects for two years and came up with disappointing results. The supplements produced no more pain relief than the placebo, and X-rays revealed no significant slowing effect on the rate at which space between bones in joints—a measure of cartilage health—was lost. (See page 70 for more on this study.)

The failure of such large clinical trials is unlikely to deter the manufacturers of chondroitin and glucosamine supplements from hyping their wares. There is always some study they can point to that shows some sort of result that can be interpreted as being positive, and thanks to the placebo effect, there is never a problem lining up impressive testimonials. Juices and foods fortified with glucosamine, sparkling with potential profit—at least for the producer—are joining glucosamine pills on store shelves. They have fancy labels and prices, but not much in terms of evidence.

So if glucosamine or chondroitin are not very effective in alleviating the symptoms of osteoarthritis, is there anything that is? Maybe. Extracts of a South African herb known as devil's claw contain the anti-inflammatory compound harpagoside and, in a few controlled trials, have allowed subjects to cut back on other pain medications. Cat's claw, a vine that grows in the Peruvian jungle, also has anti-inflammatory properties, but human trials are lacking. There is more evidence for extracts of the *Boswellia serrata* plant, commonly known as frankincense. The major active ingredient is acetyl-11-keto-beta boswellic acid, which is known to interfere with the activity of 5-lipoxygenase, an enzyme that catalyzes the formation of leukotrienes, which promote inflammation. A particular preparation known as 5-Loxin was shown to reduce knee pain and stiffness in a controlled trial. Furthermore, when fluid was drawn from the knees of subjects taking the herbal supplement, there was a significant decrease in matrix metalloproteinase-3, an enzyme that can break down cartilage.

Another plant from the Amazon, *sangre de grado* (*Croton palanostigma*), has received attention because laboratory studies have shown that an extract known as progrado contains compounds that are not only capable of interfering with the activity of metalloproteinase enzymes but also can stimulate the production of insulin-like growth factor-I (IGF-I). Chondrocytes, the cells that produce the major components of cartilage, use IGF-I in the cartilage repair process. Human trials of the extracts, which appear to be completely safe, are needed. Other natural products that—at least according to a few, often poorly controlled, studies—have provided some benefit in osteoarthritis due to their anti-inflammatory properties include sour cherry preparations, green tea extract, ginger, pine bark extract (sold under the brand name Pycnogenol) and polyphenols isolated from organically grown olives. Cetyl myristoleate, found naturally in nuts, vegetables and dairy products, is also available as a dietary supplement and appears to have an anti-inflammatory quality—interference with 5-lipoxygenase. A few human studies have shown some improvement in knee function.

Methylsulphonylmethane (MSM) is a sulphur-containing compound found in trace amounts in fruits and vegetables, and is touted as a source of sulphur, which the body needs to make critical proteins, such as those in cartilage. There is no evidence of a lack of sulphur in the North American diet, and there is no clinical evidence that it is effective for the treatment of joint pain. Horse trainers, however, claim that it keeps their animals' joints healthy. Those guys are not into throwing money away, so there may be something there. There is precious little evidence that S-adenosylmethionine (SAMe), another sulphur-containing compound that occurs naturally in the human body, helps repair cartilage and relieves pain as its promoters claim. Topical creams containing extracts of chili peppers, on the other hand, have been shown clinically to provide temporary relief. Capsaicin, the active ingredient, is known to interfere with the transmission of pain signals. And as far as acupuncture goes, take your pick from the studies that show relief of osteoarthritic pain or those that show no effect at all.

I would suggest that any benefit from apple cider vinegar, copper bracelets, magnets or homeopathic remedies that are so dilute as to contain essentially nothing comes from a mind-over-body effect. Of course, if you feel better, the why is irrelevant. As far as raisins soaked in gin go, I suppose the benefit depends on how much of the gin is consumed along with the raisins. Pump enough into the mouth and you'll forget all about your osteoarthritis.

If all of this sounds bleak, fret not. There is an increasing amount of evidence that osteoarthritis can be helped, perhaps even prevented, by something as simple as choosing the right foods. Vitamin C is needed to make both collagen and proteoglycans, the essential components of cartilage. Osteoarthritis is less prevalent in people who consume at least 150 milligrams of vitamin C a day. Vitamin C also neutralizes free radicals that are often responsible for joint inflammation. So eat your fruits and vegetables. And your vitamin D–enriched foods. The health of cartilage depends, to a degree, on the health of the underlying bone, proper formation of which requires both calcium and vitamin D. But perhaps the most important dietary factor in osteoarthritis is the composition of the fats we consume.

Fatty acids that are produced when the body breaks down fats are very active biologically. They get incorporated into cell membranes and determine their fluidity, which in turn affects how cells communicate with each other. Fatty acids are also precursors for important biochemicals known as prostaglandins, some of which encourage and others suppress inflammation. In general, omega-3 fats, found in fish, flax, canola, soy and nuts, are anti-inflammatory, while omega-6 fats, found in corn, sunflower, safflower and cottonseed oil, are pro-inflammatory. Unfortunately, most processed foods are made with omega-6 fats, with the consequence that the North American diet provides ten times as much of the pro-inflammatory fats as the anti-inflammatory fats. The optimal ratio is thought to be closer to one! Basically, this means eating less meat, fewer processed foods, more fish and at least six servings a day of fruits and vegetables. Couple that with a program of regular exercise and your joints will thank you.

Of course, that may not sound as magical as autohemotherapy. What's that? As described in the *Indian Journal of Orthopedics,* the treatment involves oral administration of a concoction made by mixing a patient's blood with honey and lemon juice in a copper bowl. My guess is that, like so many remedies touted to treat osteoarthritis, it is of no bloody use.

GROWING PROFITS

WHEN YOU WALK through the gates of the Chelsea Physic Garden in London, you enter a living pharmacy. But instead of finding shelves with bottles, you'll find beds of plants with intriguing signs like Cardiology, Parasitology or Anesthesiology. Inscriptions identify the species from which medications such as digitalis (heart disease), quinine (malaria) or colchicine (gout) are derived. This amazing garden was founded in 1673 with the purpose of training apothecary apprentices in identifying plants. Indeed, back then, botanicals were the main source of drugs, and an apothecary had to learn how to use ephedra for lung problems, mandrake for pain or licorice root for an upset stomach. Modern chemistry has allowed the active ingredients in many plants to be extracted, identified and standardized for use in the roughly one-quarter of our current prescription drugs that originate in plants. But if someone is in the throes of pain, physicians don't direct them to graze in a field of poppies; they prescribe an appropriate dose of pure morphine, isolated from the flower. Plants, though, are very complex chemically, and possible therapeutic effects cannot always be traced to a single active ingredient. In fact, in some cases it may be the synergistic action of various components that provides the benefit. That is why there is so much interest today in using some sort of standardized versions of whole herbs.

Potential for improved health is not the only reason for the herbal boom. There is also the prospect of a nifty little profit. The widespread belief that natural products are somehow inherently safer and more effective than synthetic drugs makes for a very lucrative market. The danger or efficacy of a substance does not depend on whether it comes from a bush or a lab; it depends on its molecular structure. But producers are nonetheless quick to capitalize on people's romanticized view of herbs and have flooded the market with preparations that contain more hype than active ingredients. Take, for example, *Prunella vulgaris*, a weed that has been blessed with the highly marketable common name of heal-all. Traditionally it has been used, as the name implies, to treat virtually any disease. Thyroid problems, diarrhea, sore throats, colds, liver ailments and "weakness of the womb," whatever that may mean, are just some of the conditions that it is credited with "healing." Of course, just because heal-all has been used traditionally to treat such conditions does not mean that it has been used to treat these conditions effectively. There is no scientific evidence to back up the claims, but the Canadian Natural Health Products Directorate (NHPD), which was created to solidify the herbal quagmire, does not require any. Approval to sell a product for "traditional use" requires only that the substance has an acceptable safety record for at least fifty years, and that it has been used within a "cultural belief system or healing paradigm." I suppose this is not too big a surprise, given that almost half the members of the NHPD's Expert Advisory Committee are practising herbalists.

The proliferation of nonsensical claims based on "traditional use" is troublesome because it casts a shadow on the whole herbal market. But let's not throw out the baby with the bathwater. Serious scientists are exploring the therapeutic benefits of plants and are coming up with some pretty alluring results. Evening primrose oil, for example, may eventually play a role in the treatment of some breast cancers (those described as Her-2/neu-positive). Ginkgo biloba, which has an exaggerated reputation as a memory enhancer, may actually help fend off ovarian cancer. Researchers at Brigham and Women's Hospital in Boston compared the lifestyles of six hundred women with ovarian cancer to a similar number of healthy controls and found that "4.2 per cent of ovarian cancer-free women reported taking ginkgo regularly for at least six months before diagnosis, but only 1.6 per cent of women with ovarian cancer reported taking ginkgo." Not a huge difference, but statistically significant, and made more meaningful by follow-up studies that showed that ginkgolides, compounds found in ginkgo, inhibited the growth of ovarian cancer cells in the lab. Nobody is yet recommending that women start taking ginkgo biloba supplements to ward off ovarian cancer, but future research may point in that direction.

People with hay fever may benefit from research being carried out with extracts of the leaves of the butterbur plant. Compounds called petasines have been shown to inhibit the body's production of leukotrienes, which are allergy and asthma mediators. You will note that the word "may" appears very often in discussions of herbal treatments, because hard scientific evidence is elusive. But in one interesting case, we may be able to remove the may. Maybe.

Ginseng has long been associated with a number of fanciful health benefits, including increased energy, improved mental function and a vitalization of the immune system. A Canadian company, CV Technologies of Edmonton (now known as Afexa Life Sciences), has demonstrated through sound research that there is something to the claims of improved immune function. A specific preparation of North American ginseng, containing a standardized amount of certain polysaccharides that are thought to be the active ingredients, has been shown in clinical trials to have an effect on the common cold. It isn't quite the Holy Grail, but in a proper, controlled, double-blind trial, its product Cold-fX reduced the frequency and severity of the common cold.

For four months, 323 adults took either the ginseng capsules or a daily placebo and kept careful records of any symptom that could be attributed to the common cold. Cold-fX did not prevent people from catching a cold, but it did reduce the risk of contracting a second one and it did reduce the misery commonly associated with a cold. We are not talking of a miracle here, but in spite of the fact that the product is promoted by Don Cherry, the famous hockey pundit, it is an herbal product that does work. Just goes to show that nobody can be wrong about everything all the time.

WELCOME WEED

SEAWEED IS PRIZED in some cultures for its reputed medicinal qualities. While there is little evidence for any therapeutic effect, the green stuff may just turn out to be the remedy needed to restore an ailing Canadian pharmaceutical company to good health. Bellus Health has been in the financial doldrums since 2007, when it announced that the long-awaited results of its trial of the anti-Alzheimer's drug tramiprosate (proposed trade name Alzhemed) were disappointing. This had been the company's flagship product, the one that was to generate billions of dollars in profits once approved by the U.S. Food and Drug Administration. But the drug's chances for approval vanished when the company's well-designed, large-scale study showed that Alzheimer's patients benefited no more from tramiprosate than they did from a placebo.

It now seemed that the 250 million dollars and fifteen years the company had invested in tramiprosate research were destined to go down the drain, along with investors' money. Could anything be done to salvage the drug? Given that the active ingredient is purported to occur naturally in a type of seaweed, perhaps. Although there had been insufficient evidence to convince the FDA to approve Alzhemed as a prescription medication, it could still make it to market as a "natural health product," a category of substances much more loosely regulated than prescription drugs. So Alzhemed was reinvented as Vivimind, and the advertising machinery went into high gear, promoting it as "memory protection" for healthy people. Aging baby boomers who were beginning to fear that names didn't roll off their tongues with the usual ease were ideal targets.

To be sure, there is some clever science behind tramiprosate. The drug was specifically designed to hinder the formation of amyloid plaque, deposits in the brain that are the hallmark of Alzheimer's disease. These deposits gum up the brain's machinery by interfering with the transmission of signals between nerve cells, a process that is fundamental to thinking and memory. Researchers wondered how this could be prevented. Logically, the first step was to determine the chemical composition of such plaque, a step that has been successfully taken.

The fundamental reaction involved in plaque formation links two naturally occurring soluble substances to form an insoluble aggregate. Beta amyloid peptide, a molecule consisting of a short chain of amino acids, is one of the culprits. It causes neurological mischief when it reacts with certain complex carbohydrates known as glycosaminoglycans, or GAGs. The pharmaceutical challenge, then, is to prevent the peptide from forging a union with the GAGs. To achieve this, researchers must identify the particular feature of their molecules that allows these substances to react. Once that is clarified, a drug that incorporates this feature can be designed to serve as a surrogate partner for one of the substances involved in plaque formation. Of course, the key condition is that the new union must not yield any insoluble substance that can interfere with nerve-cell activity.

Back in the 1990s, researchers, particularly at Neurochem (eventually to become Bellus Health), began to explore model compounds they hoped would bind preferentially to beta amyloid peptide and prevent it from reacting with GAGs. Chemists synthesized a number of molecules they believed had the required features, eventually homing in on a rather simple one known as homotaurine. It was soluble, it crossed the blood-brain barrier and it tied up beta amyloid peptide, nipping plaque formation in the bud. Homotaurine was rechristened tramiprosate, and the company began investigating its possible use as an anti-Alzheimer's drug.

As with most drugs, initial studies involved mice—in this case, animals that had been bred to be especially predisposed to Alzheimer's disease. They fared very well, showing reductions in plaque formation without any significant side effects. Neither did rats nor dogs exposed to higher doses suffer side effects, except for a few instances of diarrhea. Next step was to test the safety of tramiprosate in healthy human subjects. Aside from rare cases of nausea, there were no problems. With the safety of the drug now established, the time had come to test its efficacy in humans.

Fifty-eight patients with mild to moderate Alzheimer's disease were enrolled in a three-year-long study. Analysis of their cerebrospinal fluid showed that drug use resulted in reduced levels of beta amyloid protein, and tests measuring cognitive ability also suggested some modest improvement. Although one would hardly call the results spectacular, they were encouraging enough to mount a Phase III trial, critical for FDA approval, in which more than a thousand Alzheimer's patients at a number of centres across North America would be treated with tramiprosate for two and a half years. Bellus Health and its investors eagerly awaited the results. Alas, they were to be disappointed. The findings showed insufficient improvement with tramiprosate to gain FDA approval. There would be no $4.5-billion-a-year wonder drug. The company's stock plunged.

And then someone at Bellus had a bright idea. Since company researchers claimed they had found homotaurine to be present in some sort of seaweed, it qualified as a natural health product, even though the company's version was arrived at through chemical synthesis. True, this meant that Vivimind could not be marketed as a treatment for Alzheimer's, and that claims on its behalf had to be quite sedate. Statements such as "sustains brain cell health" or "protects the brain structure associated with memory and learning" were vague enough to pass muster. These claims were buttressed by a brain-scan study of subjects taking tramiprosate that was spun into a claim of "scientifically proven to help protect memory function." The study had not shown any improvement in memory, but had shown reduced shrinkage of the hippocampus, an area of the brain linked with memory.

So, can this failed prescription Alzheimer's drug reduce memory loss in healthy people, as Bellus proposes? It has never been tested for that, so we will just have to wait and see. Certainly, a fair degree of skepticism is warranted. However, given North America's current passion for all things "natural," there is a good chance for market success. After all, effective promotion often trumps evidence. Finally, I'd say that whoever at Bellus Health came up with the idea of connecting tramiprosate with seaweed deserves a raise. Otherwise, tramiprosate would probably be dead in the water.

CANCER PREVENTION CHECKLIST

IMAGINE THAT YOU had the opportunity to have any question you can think of answered—but only one. What would it be? I've got a candidate: "How can cancer be prevented?" A pretty important question, given that roughly one in two or three of us (depending on whose statistics you believe) has to cope with the dreaded disease at some time in our lives. And I'm not only proposing a question; I also have an answer. Well, it isn't really *my* answer. It comes from the most extensive and exhaustive survey of the scientific literature on cancer prevention ever undertaken.

The World Cancer Research Fund and the American Institute for Cancer Research are both organizations with no commercial ties, dedicated to helping people make choices about reducing their chance of developing cancer. Their latest collaborative effort, *Food, Nutrition, Physical Activity and the Prevention of Cancer*, is the result of years of work by a panel of experts from around the world and represents a comprehensive review of thousands of publications. The report itself is like a telephone book, but its essence can be summarized relatively easily.

The main thrust was to examine the link between foods and specific cancers, ranking the associations from "convincing decreased risk" to "convincing increased risk," with "probable decreased risk" and "probable increased risk" in between. So, what was the conclusion about convincing increased risk? Red meat and processed meat clearly increase the risk of colorectal cancer; arsenic in drinking water increases the risk of lung cancer; and alcoholic drinks increase the risks of cancers of the mouth, pharynx, larynx, colon and breast. Aflatoxins, found in mouldy grains, cause liver cancer. And body fatness increases the risk of esophageal, pancreatic, colorectal, breast, endometrial and kidney cancers. Probable increased risks include salt consumption for stomach cancer, diets high in calcium for prostate cancer, alcoholic drinks for liver cancer and arsenic in water supplies for skin cancer.

Now for convincing decreased risk. Pretty short list here. Breastfeeding exclusively for six months decreases the risk of breast cancer. It also reduces the risk of obesity in children, which in turn reduces the risk of cancer. Physical activity decreases the risk of colorectal cancer. Probable decreased risk is more encompassing. Dietary fibre and garlic both decrease the risk of colorectal cancer, and foods containing selenium or lycopene decrease the risk of prostate cancer. Milk decreases the risk of colorectal cancer, and so do calcium supplements. Foods with folate reduce the risk of pancreatic cancer. Consuming fruit and non-starchy vegetables decreases the risk of cancers of the mouth, esophagus, stomach and lungs.

There were some surprises. Smoked or barbecued animal foods were seen to increase the risk of stomach cancer only slightly, and fat consumption only affected lung and breast cancer, and not with high significance. No evidence was found that vitamin supplements decrease risk.

After digesting the massive amount of data, the expert panel managed to distill the information down to eight recommendations. Here they are:

1. Overweight increases the risk of many cancers. Aim for a body mass index (BMI) between 21 and 23. The BMI is determined by dividing body weight in kilograms by the square of the height in metres.

2. Aim for thirty minutes of vigorous or sixty minutes of moderate physical activity a day.

3. Consume "energy-dense" foods (those that contain more than 225 calories per 100 grams) sparingly. Aim for foods with less than 125 calories per 100 grams. This means very limited consumption of fast foods and sugary drinks.

4. Eat mostly foods of plant origin. Try for some unprocessed grains and/or legumes at every meal. Eat at least five servings of fruits and non-starchy vegetables a day. Non-starchy vegetables include broccoli, carrots, green leafy vegetables and bok choy, but not potatoes.

5. Limit cooked red meat to less than 500 grams (18 ounces) a week, and avoid processed meats like hot dogs, hams, salamis and smoked meat. Ouch!

6. Avoid alcoholic drinks. The panel took into account the fact that modest alcohol consumption may protect against heart disease, but found that as far as cancer is concerned, there is no level of consumption below which there is no increase in risk. If alcohol is consumed, it

      should be limited to two drinks a day for men and one
      for women.

7. Limit salt intake to less than 6 grams a day (corresponding
    to 2.4 grams of sodium). The best way to do this is to
    curb processed foods and fast foods.

8. Dietary supplements are not recommended for cancer
    prevention, but vitamin D, which holds the greatest
    hope, was not investigated. It should be added here that
    the panel did not consider the possible benefit of sup-
    plements for the prevention of other diseases and did
    not find any risks with supplements.

There you have it. The culmination of a five-year process. On first glance, it may seem as if there is not much new here, but there actually is. This is the first time that any expert group has made such specific recommendations about processed meats, soft drinks and alcoholic beverages. The advice is to avoid them. And what can we hope to achieve by adhering to these recommendations? The prevention of some 30 to 40 per cent of cancers. If we include giving up smoking, we're probably up to preventing 60 to 70 per cent of cases.

What about the rest? Well, genes do play a role and so do certain chemicals. There are close to 100,000 chemicals used in commerce today, with very few having been thoroughly evaluated in terms of human cancer risk. That's the next step. We need to get a handle on which of these compounds found in our cleaning agents, fabrics, plastics, toys, cosmetics, electronic equipment and foods may contribute to human cancer rates. But for now, there is sufficient evidence to suggest limiting beer guzzling and doing without the pepperoni on the pizza while watching TV. And less TV-watching is also in order. The survey found that television viewing was "probably" linked with weight gain. Now for the good news: you don't have to give up coffee. There was no link to any kind of cancer. Finally, don't stress yourselves too much about all this; stress impairs the immune system, which in turn increases the risk of cancer.

DON'T PANIC OVER GRAPEFRUIT

ARE WE NOW GOING TO WORRY about eating grapefruit? There just may be an issue here, abeit a small one, and of concern only to post-menapausal women. Our story begins back in 1989, when Canadian researcher Dr. David Bailey at the London Health Sciences Centre made an accidental discovery while studying the effect of alcohol on a blood pressure–lowering medication. Subjects who had been given grapefruit juice to mask the taste of alcohol were found to have higher blood levels of the drug than expected. As it turns out, grapefruit contains compounds, since identified as furanocoumarins, capable of interfering with the activity of certain enzymes the body uses in its metabolic processes. It's the fate of many medications to be decomposed by cytochrome P450 enzymes—hopefully after the drugs have carried out their therapeutic purpose. If these enzymes are inhibited, blood levels of the medications can rise to potentially dangerous levels. A number of medications have been found to be subject to this grapefruit effect, and since furanocoumarins can inhibit enzymes for up to twenty-four hours, patients are advised to avoid grapefruit juice completely.

Estrogens, as found in post-menopausal hormone-replacement products, are metabolized by cytochrome enzymes and therefore come with a package insert warning about increased blood levels if taken with grapefruit juice. Given that a risk of breast cancer is associated with estrogen concentrations, any rise in circulating estrogen levels is a concern. And this, of course, also applies to the estrogen produced naturally in the body. There is no reason to think that estrogen supplements are handled any differently in the body than naturally produced estrogen, which implies that women who consume grapefruit could have higher levels of estrogen and be potentially at greater risk for breast cancer. A plausible theory, but one in need of factual backing.

There are several ways such a hypothesis can be put to the test. A randomized controlled-intervention trial would be ideal. One set of subjects would be asked to consume grapefruit on a regular basis, while a control group with similar lifestyle factors would be asked to avoid the fruit. The groups would have to be very large to ensure statistical significance, and would have to be followed for many years. If there were more breast cancers in the grape-fruit-juice group, we would have our answer. Such a study, though, would be expensive and extremely difficult to organize and administer. But there is another way to get at a possible link between grapefruit consumption and breast cancer. Women who develop the disease can be compared to those who don't, in the hope that some dietary or lifestyle factor responsible for triggering the disease can be uncovered. Again, large numbers of subjects are required, along with information about their lifestyles. As it turns out, the Hawaii–Los Angeles Multiethnic Cohort Study, although not designed for this purpose, was just about made to order to investigate the grapefruit hypothesis.

Between 1993 and 1996, more than two hundred thousand Hawaiians and Californians between the ages of forty-five and seventy-five were mailed a very detailed questionnaire asking about food consumption, family history of cancer, medical conditions, medications taken and various aspects of personal behaviour. By 2002, a total of 1,657 cases of breast cancer had been identified from cancer registries, and it was at this point that Dr. Kristine Monroe and her group at the University of Southern California decided to investigate the grapefruit effect in this population.

The original food questionnaire had asked about the frequency of grapefruit consumption, which the researchers now sought to relate to the incidence of breast cancer. Investigating such a potential relationship presents numerous statistical challenges. Other known risk factors for breast cancer, such as age of puberty, number of children, age at menopause, alcohol consumption, weight and hormone use, as well as dietary fat and fibre intake, have to be adjusted for. With a population of this size, though, that can be done and the effect of a single food can be investigated.

Given that grapefruit compounds have been shown to inhibit the enzymes responsible for estrogen metabolism, the results found by Monroe's group were not totally surprising. Consuming grapefruit was associated with a 30 per cent greater risk of breast cancer. What does that mean? During the period of this study, 1,657 postmenopausal women out of 46,080 developed breast cancer, for an incidence rate of 3.6 per cent. A 30 per cent increase in this rate means roughly one extra case of breast cancer in every hundred grapefruit eaters. But how much grapefruit does one have to eat to increase the risk? This is where the surprise lies. The 30 per cent increase was seen with the consumption of just one quarter of a grapefruit per day! Unfortunately, the questionnaire did not ask about grapefruit juice consumption separately from all citrus juice consumption, so there is no way to tease out a grapefruit juice effect, if there is one. It is pretty reasonable, though, to assume that the juice would exert a similar effect, since it obviously does with medications that are metabolized by the same enzyme system as estrogens.

What, then, are we to do with this finding? Panic? No. One single study does not science make. And let's remember that the population studied was post-menopausal women, so it reveals nothing about a grapefruit effect in younger women. Still, given that an effect was found, and that it is biologically plausible, we can perhaps question the wisdom of regular grapefruit consumption by post-menopausal women. Of course, they can happily indulge in oranges.

LIAR, LIAR, BRAIN ON FIRE

I DOUBT that there is a magnetic resonance imager (MRI) in the vicinity of Daltry, a small village in the Scottish Highlands. But perhaps Dr. Daniel Langleben of the University of Pennsylvania should consider setting up a mobile unit there. Langleben, you see, uses a technique called functional MRI to study goings-on in the brains of liars. And there are liars aplenty in Daltry, according to the BBC's h2g2 website. Once a year they assemble to exhibit their talents in the Thistle public house and vie for the honour of being named liar of the year. They sure do tell some whoppers. One winner regaled the audience with a dramatic account of the destruction of his vocal cords during a sandstorm in the Sahara. Of course, the story was delivered in a clear, loud voice.

This is just the sort of gentleman Dr. Langleben would love to propel into his MRI machine to see whether or not the instrument calls him a liar. His research group has been steadily accumulating brain scans from volunteers who have been asked to lie or tell the truth in diverse situations, and indications are that the clever machine can detect liars with an accuracy rate of 99 per cent. But can accomplished liars beat the machine? After all, they have been known to get the best of lie detectors, which is why these instruments are not admissible in court. Functional magnetic resonance imaging, though, should be more reliable because it depends on brain activity, not on heart rate or sweating, which adept liars can learn to control.

Functional MRI is a specialized version of magnetic resonance imaging, that outstanding technique for peering into the body in a non-invasive fashion. Although MRI is highly sophisticated, the basic principles are readily understandable. Just think of the body as a collage of proteins, fats, carbohydrates, nucleic acids, water and various other chemicals, almost all of which contain hydrogen atoms. In fact, about 60 per cent of the atoms in our body are hydrogen, and each one of these has a nucleus consisting of a single

proton, a tiny positively charged particle with magnetic properties. When a patient lies within the very large magnetic field generated by the MRI machine, the hydrogen nuclei in the body tend
to line up with the magnetic field, in much the same way a com-
pass needle lines up with the Earth's magnetic field. Introducing a
pulse of radio waves imparts enough energy to cause the hydrogen
nuclei to flip around and oppose the external field. As soon as the
pulse is switched off, the nuclei "relax" and revert to their original
orientation, giving off the energy they had absorbed. The instrument can pinpoint where this energy is coming from and thus locate
the exact position of the hydrogen atoms. Since all tissues contain
hydrogen, an image of the inside of the body can now be produced.

But the real key to MRI is that the time it takes for the hydrogen
nuclei to relax after the radio pulse is switched off depends on their
specific molecular surroundings. The liver, for example, is a complex organ whose various hydrogens are in different environments
and will have different relaxation times. An MRI scan can detect
these times and, through a system of colour-coding, translate them
into an image. What we now see is a beautifully detailed, multicoloured, anatomical picture of the liver. A tumour can be readily
spotted because its hydrogens, being in a different chemical environment, will show up on the image in a different colour. Similarly,
different parts of the body can be investigated, including the brain.
And this is where functional MRI can supplement the information
available from regular MRI. It provides insight into which parts of
the brain are activated when we feel pain or joy, when we fall in
love—or when we lie.

Functional MRI makes use of the fact that brain activity is
associated with blood flow. When a part of the brain becomes
active, small blood vessels dilate to allow more oxygenated blood
to rush to the scene. The oxygen is actually delivered by binding
to iron atoms in hemoglobin molecules. Iron can cause small
distortions in magnetic fields, and the degree of distortion

depends on whether the iron atoms are carrying oxygen or not. What this means is that hydrogen nuclei in oxygenated blood are in a slightly different magnetic environment than those in deoxygenated blood. Therefore, by taking sequential scans, an appropriately tuned MRI instrument can detect which parts of the brain are active because activity requires energy, and energy production in turn requires a good supply of oxygenated blood. Since the hydrogen nuclei in oxygenated blood relax at a different rate, a compilation of scans produces an image in which active brain parts show up in a different colour.

When a subject's hands are placed in hot water, for example, a different part of the brain is highlighted than when he is shown pictures of a loved one. Similarly, joy and fear trigger responses in specific locales, as do smelling fragrances and, it seems, telling lies. Lying requires more thinking, and therefore more energy, than telling the truth. This is where Daniel Langleben's fascinating research comes in. Volunteers were placed in an MRI scanner and were asked to view a card that flashed on a screen in front of them. They were then to lie or tell the truth about whether the card matched one they had been previously shown. The researchers were able to detect lies with such an amazing degree of accuracy that they now suggest the technique can be useful in criminal and even terrorist investigations. But what isn't clear is whether liars can learn to fool functional MRI. That is why participants in Daltry's Liars' Contest would be great subjects to study. Just imagine looking into the brain of Fraser Patrick McInnon, who won his eight consecutive title with the immortal words, "I regret I cannot enter the contest this year, as I cannot tell a lie."

Unfortunately, however, Mr. McInnon may be hard to find, since on further investigation it seems the story found on the BBC's website may itself be a clever lie. But I can assure you that the story about detecting lies by functional magnetic resonance imagining is true.

Of course, you would have to see my brain scan to be sure of that.

JUST THE FACTS

IN 1828, Friedrich Wöhler, a professor of chemistry at the University of Göttingen, excitedly wrote to Jakob Berzelius, his former mentor, "I have witnessed the great tragedy of science, the slaying of a beautiful hypothesis by an ugly fact." Wöhler was referring to his synthesis in the lab of urea, an organic compound that previously had only been isolated from living systems. The prevailing opinion at the time was that such substances contained a "vital force" that could not be reproduced by man. An interesting theory, but as Wöhler showed, not borne out by facts. Scientific progress must, of course, be built upon facts, and if these slay a pet theory, well, so be it.

Recently, several studies have come to light that may not exactly have slain widely held theories about health, but have surely inflicted some pretty serious wounds. Take, for example, the popular belief that glucosamine and or chondroitin can significantly reduce the pain of osteoarthritis. Glucosamine, derived from crab shells, and chondroitin, from animal cartilage, are widely sold as alternative treatments for osteoarthritis, the age-related condition that can cause severe pain as the bones' protective layer of cartilage erodes and bone rubs on bone. There is a chemical similarity between the composition of cartilage and that of glucosamine and chondroitin, so it is tempting to conclude that these may have some therapeutic value. Physiologically, it seems unlikely that oral doses of these substances can survive digestion and find their way to where they are needed to restore cartilage. But given the plethora of anecdotal reports about pain reduction with dietary glucosamine and chondroitin, the National Center for Complementary and Alternative Medicine in the U.S. decided to carry out a rigorous study.

In the Glucosamine/Chondroitin Arthritis Intervention Trial, more than fifteen hundred patients across the U.S. were given glucosamine, chondroitin, a combination of the two or a placebo.

The doses were the ones commonly recommended by supplement manufacturers, namely 500 milligrams of glucosamine and 400 milligrams of chondroitin taken three times a day. The pills used had to be specially formulated for the trial because none of the commercial products met the FDA's requirement for dose reproducibility. Unfortunately, after twenty-four weeks, the results were not very encouraging. Although the researchers looked only for a 20 per cent improvement in pain as measured by a standard test known as the Western Ontario and McMaster Universities Osteoarthritis Index, they didn't find it. Glucosamine, chondroitin or the combination performed no better than the placebo. There was, however, statistically significant improvement in a small subset of patients who had severe pain.

Needless to say, the study has been criticized. Glucosamine proponents point to a European study, published at about the same time, that compared glucosamine with acetaminophen (Tylenol) and placebo in 318 patients. In this double-blind trial, glucosamine significantly reduced pain, but acetaminophen did not. Why the different findings? Well, the Europeans used glucosamine sulphate, the Americans treated with glucosamine hydrochloride. Hard to imagine that this makes a difference, but maybe it does. It is more likely, though, that the patients in the European study had more pain and the supplement works better in such patients. So what is the bottom line? Unless the pain of osteoarthritis is very severe, neither glucosamine nor chondroitin is helpful. For severe pain, the combination is worth a try, but remember that the researchers considered a 20 per cent improvement in pain level a success! Unfortunately, the solution to osteoarthritis does not seem to lie in guzzling pills made from animal cartilage or crab shells. And neither does the solution to heart disease lie in popping vitamin B supplements.

That was a good theory too. It first appeared in the late 1960s when Dr. Kilmer McCully noted that children born with an inherited disease that raised levels of the naturally occurring compound homocysteine in their blood had an increased risk of premature coronary disease and stroke. Since the normal breakdown of homocysteine requires the presence of B vitamins, it was seductive to conclude that vitamin supplements could reduce the risk of heart disease. Many doctors started measuring homocysteine in patients' blood and recommended vitamin B supplements if the levels were elevated. As predicted, homocysteine levels did decline in response to the supplements, but the question of whether this led to a decline in risk of heart disease was still open. Until now. It seems we may have been beating the wrong bush.

The Heart Outcomes Project Evaluation (HOPE) was designed to determine the effect of lowering homocysteine on heart attack or stroke. More than five thousand patients at risk due to existing vascular disease or diabetes were given B vitamins or a placebo. After five years, the subjects who took 2.5 milligrams of folic acid, 50 milligrams of vitamin B_6 and 1 milligram of vitamin B_{12} daily were no better off than those who took a placebo. This, in spite of a 25 per cent reduction in blood homocysteine.

A Norwegian study that involved giving B vitamins to men and women after a heart attack came to the same conclusion. Again, homocysteine was reduced, but the risk of a second heart attack or sudden death was not. Homocysteine, it seems, may signal the approach of heart·disease, but doesn't cause it. High doses of vitamin B supplements to reduce the risk of heart disease is therefore not warranted. But adding folic acid to flour may be a different story. Since this measure was introduced with the goal of reducing the incidence of birth defects (which it did successfully), the number of deaths from stroke in North America has dropped significantly. In England, where flour has not been fortified, the incidence of stroke has not decreased. So, where else can you get folic acid besides flour? All sorts of leafy green vegetables, as well as asparagus, beans and chick peas. Load up on these. A lot more evidence of benefit here than with the vitamin B pills. And for osteoarthritis? Well, at least one laboratory study has shown that pomegranate extract can reduce the damage to cartilage seen in osteoarthritis. We'll have to wait to see if some ugly fact emerges to slay this beautiful theory.

DANGER! DO VACCINATE

AVOID! One of the most common words of advice heard these days. Avoid tap water. Avoid bottled water. Avoid butter. Avoid margarine. Avoid the sun. Avoid sweeteners. Avoid genetically modified foods. Avoid plastic bags. Avoid paper bags. Avoid preservatives. Avoid dairy. Avoid meat. Avoid soy. Avoid—ah, never mind. I could go on and on with a litany of such "avoids." There are some valid points to be made with some of these, but there is one avoid that I cannot stomach. Advising parents to avoid childhood vaccination is scientifically unjustified and dangerous.

Vaccination just may be the most significant medical advance in history. It is difficult to estimate the number of lives saved, but it is in the many millions, to say nothing of the countless number of people who have been spared the misery of mumps, measles, whooping cough and polio. I can vouch for the agony of whooping cough myself. Feeling as if you are going to cough your lungs out is a memory that doesn't leave you easily. I survived, but one of my classmates in grade two did not. And how often can one say that a disease has been completely wiped off the face of the Earth by a medical intervention? The last case of smallpox was recorded in 1978. The World Health Organization estimates that smallpox killed as many as 500 million people in the twentieth century and that as recently as 1967 it was responsible for two million annual deaths.

Other vaccines may not have eradicated diseases, but they have curbed their incidence very significantly. Cases of whooping cough in North America have declined from a pre-vaccination peak rate of about 300,000 per year to 10,000. Measles from a million cases a year to a hundred. Diphtheria and polio are almost nonexistent today in developed countries. The incidence of hepatitis B and tetanus have been reduced by a factor of forty, rubella by two hundred and mumps by four hundred. The effectiveness of immunization is simply beyond argument. How can there be an issue here? How can some parents choose not to vaccinate their children?

It really is a conundrum. But the answer likely lies in a growing distrust of the "medical establishment," a discredited but widely publicized scientific study, inaccurate information being spread on the Internet, and a lack of understanding of the difference between an association and a cause-and-effect relationship.

Although we may not think of it in such terms, the decisions we make in life often come down to a risk-benefit analysis. Whether it is flying in airplanes, eating smoked meat, taking cholesterol-lowering medication, or vaccination, there are always pluses and minuses to consider. There is no denying that immunization does come with some risk. Rashes, joint pain and fever are well documented, as are occasional lapses in the speed with which safety issues concerning vaccines have been addressed. Oral polio vaccines, which were more convenient to administer than the injected form, were responsible for actually causing the disease in rare cases. Yet some twenty years were allowed to pass before switching back to the safer, injectable form. An infant vaccine against an intestinal infection that struck roughly four million babies a year in North America was found to cause an increase in life-threatening cases of bowel collapse and had to be abandoned. Although there is no scientific evidence linking the mercury-containing preservative thimerosal to any disease, it probably should have been removed from vaccines more speedily when ill effects attributed to mercury in other contexts became apparent.

Vaccines, in a sense, are becoming victims of their own success. As memories fade of the horrors of the original diseases that they prevent, more attention is being focused on possible harmful side effects. Indeed, one can judge the progress of society by looking at its worries. Instead of having to be concerned about millions dying from smallpox or coming down with measles or whooping cough, we worry about the possibility of vaccination being linked with some cases of autism. That suggestion was raised in 1998 by a paper published in the British medical journal *The Lancet.* Andrew Wakefield and twelve colleagues claimed that the measles, mumps and rubella vaccine (MMR) caused a bowel disease that then caused autism.

The report received extensive publicity and triggered public demonstrations against mandatory vaccination. Most scientists were skeptical of the Wakefield study, and their skepticism was borne out by the results of an investigation published in 2002 in the *New England Journal of Medicine.* Danish researchers had examined immunization records and autism diagnoses for all children born between 1991 and 1998 and found that unvaccinated children were just as likely to be diagnosed with autism as those who had received immunizations. The *Lancet* study was further discredited when it was revealed that Wakefield had failed to disclose he had received a large grant from a group of lawyers who were looking for ammunition in a lawsuit against vaccine manufacturers. In the end, ten of Wakefield's co-authors retracted their support for the original research, saying that in retrospect the results as reported were not valid.

Other studies around the world also refuted the link between vaccines and autism, but a vocal group of anti-vaccine advocates maintains that a witch hunt has been organized against Wakefield to protect vaccination interests. Humbug. The fact is that autism commonly shows up at roughly the same age that vaccines are given, and an association can readily be mistaken for a cause-and-effect relationship. But even if there really were a link between autism and vaccination, the anti-vaccine movement would still not be justified. The benefits overwhelm the risks.

In Britain, the consequences of the vaccine scare are already being seen in rising rates of mumps, rubella and measles. And Britain faces another problem: homeopaths are recommending that tourists travelling to malaria-stricken destinations use homeo-pathic remedies instead of well-tested prescription prophylaxis. This is ludicrous. Homeopathic products contain no active ingre-dient of any kind, so it comes as no surprise that a number of travellers have already paid for their folly with their health. Many homeopaths also advise their patients to avoid vaccines in lieu of a cacophony of implausible homeopathic medications. If indeed you are looking for something to avoid, how about this silly and dangerous advice?

SCIENCE
WITH A
DOSE OF
NONSENSE

DIESEL'S VISION

BEARS, IT SEEMS, are not particularly fond of french fries. We know this because of a "bear attractant analysis" conducted by researchers at Washington State University's Bear Research Conservation and Education Facility on behalf of Yellowstone National Park. The ecologically minded administration wanted to switch the fuel used by the park's diesel-powered vehicles to biodiesel, made from used cooking oil. There was concern that bears might connect the strong french-fry scent of this fuel to a food reward and be attracted to the vehicles and their human contents. Luckily, the bears found the scent unappealing, and Yellowstone Park vehicles now happily cruise around wafting compounds like methanethiol, 2,3-diethyl-5-methylpyrazine and (Z)-2-nonenal into the air. These components of french-fry fragrance may sound a little ominous, but they are of less concern than the compounds and soot produced by petroleum-based diesel engines. Indeed, a switch to biodiesel reduces potential health risks. And on top of that, biodiesel does not lead to an increase in atmospheric carbon dioxide, it is biodegradable if spilled, and unlike petroleum, it is a renewable resource. All right then, so what is biodiesel?

Let's go back for a moment to the year 1900. The scene is the Paris World's Fair, where one of the main attractions is Rudolf Diesel's amazing invention. Spectators marvelled at the engine that required no spark to ignite the fuel. They had seen internal-combustion engines before, but not like this one. Indeed, some fifteen years earlier Gottlieb Daimler had already built a four-wheeled motor vehicle, and by 1886 Karl Benz had patented the gasoline-fuelled car. But in a gasoline engine the fuel, a complex mix of hydrocarbons derived from petroleum, has to be vapourized before being mixed with air and set ablaze by a spark.

Large molecules do not vapourize readily, so gasoline is made of that fraction of petroleum that contains compounds having five to ten carbon atoms. Diesel was bent on designing an engine capable of using a greater variety of fuels. He knew that when air is quickly compressed, its temperature rises dramatically, making it hot enough to ignite any fuel that is then introduced. No spark is needed, and the fuel does not have to vapourize as in a gasoline engine, meaning that thicker fuels having ten to twenty carbon atoms in their molecules can be used. Components of petroleum that could not be used in a gasoline engine burned readily in Diesel's engine—and, remarkably, so did oils derived from vegetable sources. The engine that so captivated spectators at the Paris World's Fair actually ran on peanut oil. Indeed, Rudolf Diesel was a visionary: in 1912 he made the prophetic statement that in time the use of vegetable oils for engine fuels would be as important as the use of petroleum. Well, that time is finally approaching.

Peanut oil was not an ideal fuel for the diesel engine. It worked all right, but not for very long. The oil was too viscous and the engine gummed up. Hydrocarbon fractions from petroleum worked much better and could be produced more economically than gasoline. Diesel engines also proved to be sturdier and more fuel-efficient than gasoline engines, and by the 1930s the trucks, buses and military vehicles they powered became part of the landscape. So, too, did the soot and exhaust gases they emitted. The problem wasn't only cosmetic; researchers soon began to link health problems to diesel exhaust. Much of the soot consisted of particles less than one micron in size (ultrafine particulate matter), capable of penetrating the cells of the lungs and exacerbating conditions such as asthma and bronchitis. The tiny particles also were found to cause inflammation in the lungs, making gas exchange more difficult and putting a strain on the heart. But of even greater concern was that soot particles, which are far more abundant in diesel than in gasoline engines, carried such nasties as polyaromatic hydrocarbons (PAHs), known carcinogens, deep into the lungs. The Clean Air Task Force in the U.S. estimates that diesel fumes are responsible for some twenty-five thousand deaths in North America every year.

Studies have even shown that diesel exhaust may impair the body's ability to fight off infection and that it contains compounds that, at least in pregnant test animals, may have an inappropriate masculinizing effect on the fetus. Especially worrisome are the large numbers of diesel school buses that expose children to their exhaust. Measurements have shown that sometimes the air inside the bus can be more polluted than the surroundings.

Buses fuelled by natural gas or propane run far cleaner, but biodiesel also presents a partial solution to diesel pollution. Treating vegetable oils with methanol in the presence of a sodium hydroxide catalyst produces compounds known as methyl esters, which burn very efficiently in a diesel engine and produce far less pollution than fuel made from petroleum. Used frying oil from restaurants can be converted to biodiesel, as can soy oil, rapeseed oil (also known as canola) or even fish oil. An added advantage here is that these, unlike petroleum, are all renewable resources. We can grow fuel! And although these oils do produce carbon dioxide when burned, the plants that originally synthesized them absorbed carbon dioxide from the air, meaning that with the use of biodiesel there is no net increase in atmospheric carbon dioxide. So far, we can't produce enough biodiesel to replace all diesel fuels, but it can be readily blended with petroleum-derived diesel.

Rudolf Diesel would certainly be pleased to see buses running on soy oil and would have been delighted by the escapades of Joshua and Kaia Tickell, who crossed the U.S. in their Veggie Van powered by biodiesel made from used frying oil they collected along the way. An effective way to show that alternative fuels do have a practical application! The smell of french fries attracted many a spectator, but as we saw, the Tickells did not have to worry about becoming snacks for bears.

HOT METAL

IF ONLY iron could hold on to its electrons better. We wouldn't have to worry about our cars and ships rusting or overpasses collapsing. Washing machines wouldn't leave rust stains on our clothes, tools would last longer and water pipes would spring fewer leaks.

Iron is the most widely used metal in the world, mostly in the form of steel, in which it is alloyed with up to 2 per cent carbon. Addition of carbon increases the hardness of the metal by preventing iron atoms from sliding past each other. Since our lives rely so heavily on iron, its corrosion represents an immense problem and an immense cost. Every year in North America, we spend some 500 billion dollars, close to 5 per cent of the gross national product, replacing rusted metal—all because iron is too loose with its electrons.

In nature, iron is found mostly in the form of iron oxide ore, which essentially is rust. To produce metallic iron, the ore has to be stripped of its oxygen content. This is done by heating it in a blast furnace with carbon, which combines with oxygen in the ore to form carbon dioxide, leaving a residue of iron. Essentially, carbon donates electrons to iron, "reducing" it from its positively charged ionic form in the ore to its neutral form in the metal. Corrosion is the reverse of this process, as the iron surrenders its electrons to oxygen, forming reddish iron oxide, which then flakes off the surface. More iron is exposed, forming more rust, until eventually all the metal corrodes. Rust to rust, as it were. Since energy had to be invested in the first place to separate iron from oxygen, it should come as no surprise that iron has a tendency to spontaneously revert to its oxide.

Oxygen, which makes up about 20 per cent of air, is obviously needed for rust formation. But there is another essential requirement: water. Electrons are not transferred from iron to oxygen directly—they travel though water. No water, no rust. That's why airplanes that are not in use are stored in the Arizona desert. If electrolytes (charge carriers) such as salt are present, electrons are conducted through water even more effectively. This explains why cars rust more readily in climates where salt is spread on roads to melt ice. Acids are also electrolytes, and acid rain therefore contributes significantly to corrosion. Cars will rust more quickly in areas where rainwater contains sulphuric acid, formed when industry spews sulphur dioxide into the air.

The best protection against rust is to keep moisture away from iron. Paint can do this very effectively. But there is another way to protect iron, one that makes use of ingenious chemistry. As we have seen, the corrosion of iron involves satisfying oxygen's hunger for electrons. What if this hunger could be satisfied by "sacrificing" another metal, one that gives up electrons even more readily than iron? This can actually be done in several ways. Galvanizing is a process whereby iron is coated with a thin layer of zinc, a metal that has even more trouble hanging onto its electrons than iron. In the presence of oxygen, zinc will be preferentially oxidized, forming a layer of zinc oxide. This oxide further protects the iron underneath it by blocking contact with oxygen. Chromium can be used in the same way. When alloyed with steel, it forms a protective layer of chromium oxide, giving us stainless steel.

Cathodic protection is another way to prevent iron from rusting. This involves connecting iron to a block of magnesium or zinc, both of which give up electrons more readily than iron. Oxygen will still steal electrons from the surface of the iron, but these electrons will be immediately replaced by ones from the sacrificial magnesium or zinc. The end result is that atoms of the more active metal, rather than those of iron, are converted into positive ions. There's one little catch, though: there has to be a path of moisture between the iron and the more active metal so that the oxygen, with its extra electrons, can find its way to the active metal and combine with the newly formed positive ions to form oxides. That is why ships can be protected by attaching magnesium blocks to the keel and underground oil tanks can be prevented from rusting by connecting them to an active metal. But you can't protect your car from rusting with a piece of magnesium—unless you submerge the car in water.

What happens if iron is connected to a less active metal? Then it becomes the sacrificial metal as it protects the other metal from corroding. This is why the Statue of Liberty had to be repaired in the 1990s. When it was built, its outer copper skin was supported by two thousand iron bars. Since iron is more active than copper, it preferentially gave up its electrons to oxygen, rusting away in the process. All the supports had to be replaced because the original designers were not familiar with electrochemistry. This same problem is encountered if iron water pipes are connected to copper plumbing.

Perhaps the most unusual rusting reaction has been seen in some cannonballs that have been salvaged after several centuries of sitting at the bottom of the sea. Corrosion there had caused the balls to be permeated with thousands of little holes. When pulled out of the water, the huge surface area of iron was now free to react with oxygen to produce rust, and a great deal of heat. In one instance, enough heat was generated to make a cannonball glow red.

There may not be any use for glowing cannonballs, but the principle here can be put to use if you're a skier with cold hands. Those little porous packets inserted into a glove to warm up the hands contain iron powder and water. Once the packet is removed from its plastic pouch, air enters, and the rusting process starts, generating heat. In this case, those loose iron electrons are quite handy.

COUNTERFEIT DRUGS

HUNDREDS OF LIVES could have been saved if only someone at a government lab in Panama had bothered to inject a few microlitres of the sugar-free cough syrup into a gas chromatograph. The instrument would have readily revealed that the substance destined to be used as a solvent in a government-produced cough medication was not the desired and safe glycerin, but highly toxic diethylene glycol. Immense misery and grief would have been avoided had this simple procedure, familiar even to beginning organic chemistry students, been carried out.

Glycerin is a sweet-tasting liquid that serves as an excellent solvent for the ingredients used in a number of cough, cold and fever remedies and has a well-established safety record. Actually, it is a normal product of fat metabolism, so our bodies certainly know how to handle it. Diethylene glycol, on the other hand, is a different story. This industrial chemical derived from petroleum is used in the manufacture of products ranging from coolants and plastics to inks and glues. There is no problem here, but should diethylene glycol be ingested, it can quickly destroy the kidneys and kill. Obviously, nobody would knowingly ingest such a toxic substance. But unfortunately, unknowingly, it has repeatedly happened. Diethylene glycol has solvent and sweetening properties similar to glycerin, but it costs about one-third as much to produce! Need we say more about how so many people came to ingest it? Just follow the money, as they say.

The trail leads to China, a country with huge chemical production capabilities and lax regulations. It seems that making diethylene glycol from ethylene and passing it off as glycerin doesn't present much of a problem for some chemical producers. And there appears to be no scarcity of European distributors who are content not to ask too many questions if the price is right. Certificates of analysis are forged, shipping documents are altered to hide the product's Chinese origin, complacent drug manufacturers are taken in, and bodies start piling up in the morgue. Trying to track down the companies that sell diethylene glycol as glycerin is like trying to capture water in a sieve. Chinese officials have not been as cooperative as the situation demands, unwilling to admit that companies, including some state-sponsored ones, have been complicit in the death of innocent people around the world.

But the counterfeit drug business extends far beyond substituting diethylene glycol for glycerin. It is hard to estimate the disastrous effects of infiltrating the pharmaceutical marketplace with antihypertensives, cholesterol-lowering medications, cancer drugs and malaria treatments that look like the genuine article but contain little or no active ingredient. International prescription drug counterfeiting is a rapidly growing industry, estimated to generate some $75 billion in revenues by 2010. Roughly 10 per cent of the world's medicine supply is now fake, with hundreds of thousands dying every year due to counterfeit or substandard medicine. While the majority of these cases are in Asia and Africa, the problem is emerging in North America as well.

U.S. customs officials have seized shipments of bogus Tamiflu, the antiviral drug being stockpiled to counter a possible bird flu or swine flu outbreak, which contained no active ingredient. A pharmacist in Hamilton, Ontario, was charged with dispensing counterfeit drugs after a sample of Norvasc, a medication to control hypertension and angina, was found not to be authentic. Investigation revealed that five people who had had their prescriptions filled by the pharmacist subsequently died of a heart attack or stroke.

Back in 2003, some six hundred thousand doses of fake Lipitor, the world's best-selling cholesterol-lowering medication, turned up in U.S. drugstores. Even more ominous was the recent discovery of counterfeit Procrit being unwittingly sold by reputable pharmacies to cancer patients. Procrit increases red blood cell production and is a valuable drug in the treatment of fatigue and anemia that accompanies chemotherapy, giving patients the strength to carry on with their lives. But Procrit is expensive, at five hundred dollars a dose, literally worth its weight in gold. Obviously, there is great appeal in producing a bogus version that to any observer looks like the real thing but has no costly active ingredient. And it's happening.

So far in North America, the problem has been relatively minor, but in Southeast Asia and parts of Africa it is extensive. Malaria is endemic there, and drugs in the arteminisin family are widely used to control the dreadful disease. Surveys have shown that in some areas, up to 50 per cent of the antimalarials purchased randomly from pharmacies are fake. Some contain nothing more than starch, while others have small amounts of arteminisin or its derivatives to fool testers. This is worse than having no active ingredient, because the small amounts of the drug, while not enough to kill the malaria-causing parasite, are enough to promote the evolution of resistant parasites. The consequences for millions of African malaria sufferers could be catastrophic.

Dr. Dora Akunyili, Nigeria's former national drug regulator, put it very succinctly: "Drug counterfeiting is one of the greatest atrocities of our time; it is mass murder; it is a form of terrorism against public health as well as an act of economic sabotage. It violates the right to life of innocent victims." The fake-drug racketeers do not take such verbal abuse lying down. Dr. Akunyili's office was fire-bombed and her car raked with machine gun fire, killing her driver.

Clearly, governments and the legitimate pharmaceutical industry have to take drastic steps to ensure the safety of the world's drug supply. As far as diethylene glycol goes, we've known about its toxicity since 1937, when more than a hundred people died after taking Elixir Sulfanilamide, an antibacterial medication. At the time, there was no legal requirement to test solvents used in such products for safety, but today a simple gas chromatographic analysis can quickly identify the presence of diethylene glycol. Yet, this wasn't done by the distributors of the cough medication in Panama, despite the well-documented deaths of hundreds of Haitians from a similarly adulterated product ten years earlier, and other such episodes in Argentina, India and China. As philosopher George Santayana so aptly put it, "Those who do not learn from history are doomed to repeat it."

CREAMY DREAMY

PHYSICIST, HEAL THYSELF! And he did. Mind you, it took him twelve years and six thousand experiments. But eventually, the story goes, NASA physicist Max Huber came up with a cream that healed the chemical burns he suffered in a laboratory accident. Where physicians had failed, a physicist triumphed. Better yet, he even managed to trump the chemists who had repeatedly failed to reproduce the magic formula, even after learning its components.

The clever physicist, it seemed, had managed to incorporate some magical property into his cream that defied chemical analysis. It was now time to leave rocket science behind and enter the world of cosmetic marketing. Huber's Crème de la Mer quickly achieved a cult following among the select few who could afford to plunk down hundreds of dollars for a jar, dreaming of a wondrous, rejuvenated complexion. And the competition began dreaming, too—dreaming of copying the formula and improving the complexion of their profit margins.

For chemists at cosmetics giant Estée Lauder, a company bent on reproducing Huber's cream, the dream became a nightmare. Chemical analysis revealed the components, but experiment after experiment failed to reproduce the cream. At least, so goes the tale.

I'm not sure exactly what the Estée Lauder chemists were trying to reproduce. The cream's ability to heal burns? I can't find any evidence that Huber's original formulation did that, except for his personal anecdote. No before-and-after pictures of Huber exist, nor is there any documentation of the six thousand experiments he supposedly carried out before he hit upon his secret. A rocket scientist who doesn't document his work? Very strange. When Max Huber passed away in 1991, his daughter inherited the business and sold out to Estée Lauder.

From this point on, all we have is the company's account of the story. Lauder had made a sizable investment, but just what had it bought? There was no written formula. The recipe had apparently been handed down by word of mouth and contained pretty common ingredients—mineral oil, glycerin, various plant extracts, vitamins and minerals were to be found in many similar products. Even sea kelp, on which Huber placed a great deal of emphasis, was not a novel idea, although Huber's insistence that it be picked off the northern coast of California at two specified times of the year raised a few eyebrows. In any case, the Estée Lauder chemists now had the cream's exact formula and went into the lab to duplicate it. But they could not. Batch after batch failed to yield the "activity" of Huber's original, although it isn't clear what that mysterious activity was.

Attention now shifted from the nature of the ingredients to the process that was used to make the cream. Huber's recipe had called for the ingredients to be combined and allowed to ferment for months. This means that the naturally occurring bacteria, fungi and yeasts in the sea kelp were allowed to proliferate and crank out enzymes that split the complex organic molecules in the brew into simpler substances. Fermentation, of course, is the key to making wine, pickles and penicillin, but it seemed out of place for a skin cream. Still, the Estée Lauder chemists scrupulously reproduced Huber's fermentation, but to no avail. The sought-after activity remained elusive. But there was something else in Huber's instructions that had been passed off by the serious chemists as spurious. Huber had insisted that the fermentation be carried out in a glass vessel equipped with copper plates that could be made to vibrate in response to pre-recorded gurgling sounds made by previous fermenting batches. He didn't say it had to be done by the full moon in the presence of eye of newt and toe of frog, but as far as the chemists were concerned, he might as well have. It made no scientific sense.

Nevertheless, since nothing else seemed to work, the chemists decided to let a fermenting batch listen to its ancestral sounds. And much to their surprise, the "miracle broth" now yielded the elusive activity, which judging by Crème de la Mer's promotional material, has to do with reducing the signs of skin aging. But if this special fermentation is responsible for the benefits, then it must be the result of the production of some specific compounds. And if this is the case, why are they not listed along with the other ingredients?

Of course, most people don't care about the curious chemistry that seems to be involved; they just want to know if the cream really works. La Mer insinuates that it does. "Do you believe in miracles?"

the company coyly asks in its ads. Judging by sales, some obviously do. No doubt, shelling out eighteen hundred dollars for five hundred millilitres of a face cream does motivate one to see results that may or may not be there. I am unaware of any controlled trials that have evaluated Crème de la Mer against other creams, and my guess is that Estée Lauder isn't interested in designing any. I suspect they have seen the surveys carried out by organizations such as Consumers Union, which have repeatedly shown no significant difference in satisfaction rating between regular creams and "premium" products.

La Mer's researchers are now seeking the next "breakthrough" in cream technology, namely "cosmetically smart water." Already on board is a Romanian physicist, Vasile Manzatu (yup, another physicist muscling in on the cosmetic chemists' territory), who claims to have developed a method of enhancing water's skin-penetrating ability by passing it through various electric and magnetic fields. Aging skin loses its ability to retain water, but apparently smart water can replace the water that was dumb enough to leave the safe confines of the skin. It seems another miracle is in the making. La Mer is going to turn water into gold.

Take a few dreams, add a dash of hope and toss in a measure of science. Blend in a dose of hype and pour into a fancy jar. And then advertise like mad! Make sure to throw in plenty of terms like "biopeptides," "glycosaminoglycans," "antioxidants," "SPF factors," "DNA protection" and "clinical trials." What do you have? The makings of a successful face cream.

SO, DOES ANYTHING FIX WRINKLES?

LET'S FACE IT, nobody likes the prospect of aging. And is there a more emphatic reminder that we're doing it than the appearance of wrinkles? Reducing them may not exactly turn back the clock, but it makes looking in the mirror a less disturbing experience. But how do we go about trying to iron out those wrinkles? Obviously, the first step, at least if we are going to take a scientific approach, is to have an idea of what causes them.

To put it as simply as possible, wrinkles form when molecules responsible for keeping the skin taut begin to break down. The most important of these are collagen and elastin, proteins that provide the "scaffolding" for the skin. But a major role is also played by glycosaminoglycans (GAGs) and proteoglycans (PGs), complex molecules found in the fluid between skin cells, capable of retaining large amounts of water and "plumping up" the skin. We encountered these molecules previously in our discussion of their role in the maintenance of cartilage. As we age, the cells responsible for churning out GAGs and PGs become less efficient and the skin begins to sag. The problem is compounded by aging cells producing an increasing number of those troublesome molecular species we call free radicals. These by-products of metabolism are very reactive and can damage proteins, cell membranes and even DNA. Such damage can cause the release of chemicals that cause inflammation in the skin, another factor contributing to wrinkles. Based on these observations, several approaches to designing anti-aging creams present themselves. Stimulating the production of collagen, elastin, glycosaminoglycans and proteoglycans would clearly be of benefit, as would neutralizing free radicals with antioxidants.

The first substance ever found to boost collagen production was retinoic acid, a form of vitamin A. Its use can be traced back to the 1930s, when researchers discovered that a deficiency of vitamin A in the diet resulted in skin problems. Perhaps then, the thinking went, vitamin A would also prove to be useful in the treatment of skin disorders. Good thinking! Topical retinoic acid (tretinoin) turned out to be an effective remedy for acne. Then, in the 1980s, Dr. Albert Kligman, an American dermatologist regarded as the father of retinoic acid therapy, made an interesting observation.

Patients being treated for acne saw an improvement in their wrinkles, an improvement that was eventually traced to enhanced collagen formation. But there were hurdles on the road to smooth skin. Retinoic acid treatment resulted in increased sensitivity to sunlight and, initially, in irritated skin. To ensure that patients were properly monitored, retinoic acid was made available only by prescription. But cosmetic producers are free to include other forms of vitamin A, such as retinol, in face creams. At the concentrations used, however, the effect of retinol is minimal.

Pentapeptides, now heavily marketed as anti-wrinkle agents, emerged from a study on wound healing. Researchers investigating the nature of the chemicals used by cells to trigger the formation of collagen identified these messengers as short chains of five amino acids—which puts the penta in pentapeptide. Laboratory studies of skin tissue soon showed that collagen formation could indeed be enhanced, but there was a question of pentapeptide absorption through the skin. Linking the pentapeptides to a fatty acid such as palmitic acid yielded absorbable palmitoyl pentapeptide, the "active" ingredient in many an anti-wrinkle cream. The claim is that constant use over twelve weeks results in firmer, more youthful skin. Maybe so—if you look with a microscope.

How about those glycosaminoglycans, which help the skin retain moisture and resilience? Can their production be increased? Apparently, xylose, a simple sugar, can stimulate the production of GAGs. The problem is delivering it to the appropriate cells. Just adding xylose to creams won't do, but xylose-precursor compounds that can be absorbed from creams have been developed. Once inside cells, these yield xylose, which in turn fires up GAG formation. This is not just theory: skin biopsies actually show increased amounts of GAGs. But does this translate to improved appearance? In a double-blind clinical trial, treatment with a xylose-releasing cream resulted in better skin elasticity, less dryness, fewer age spots and reduced wrinkles. The problem, though, is that these measurements are made by technicians using sophisticated instruments. Whether the difference is going to cause heads to turn is debatable.

Also debatable is the effectiveness of antioxidants in creams. In theory, they should do something. In practice, there is no compelling evidence that they do. The newest players in the antioxidant sweepstakes are nanoparticles composed of fullerenes, molecules made of sixty carbon atoms joined together in the shape of a soccer ball. Indeed, fullerenes are effective at neutralizing free radicals, but safety questions have been raised about what nanoparticles may do when absorbed into the body. Probably as much as they do in a cream.

Not a particularly flattering picture of products that suck more than a billion dollars out of North American pockets a year, is it? What does all of this come down to? Forget the hype, take the scientific lingo with a grain of salt, and find a cream that you like because of the way it feels on your face. Just make sure it has some sun protection. But face the facts. In the Consumers Union study, the best products reduced the depth of wrinkles by less than 10 per cent. You may only be able to spot this if you look into the mirror with a magnifying glass. And one last point: women's opinions about how well a cream worked were unrelated to objective measurements. Or to price. But hey, who can put a price on dreams?

GREENING DRY CLEANING

DRY CLEANING ISN'T DRY, and when it comes to environmental concerns, it isn't clean, either. But the industry is set to reform itself. To appreciate what the future may hold, a bit of history is in order. And it all starts with a clumsy maid and an alert dye manufacturer.

As the story goes, one day in 1855, Jean-Baptiste Jolly's maid knocked over a kerosene lamp on a dinner table and proceeded to mop up the spilled liquid with a cloth. When the man of the house sat down to dinner, he made a startling discovery: a segment of the tablecloth was much cleaner than the rest of the fabric, and stains that had defied laundering with water had now vanished! Asked about what had happened, the maid sheepishly admitted accidentally soaking the tablecloth with kerosene. Far from being angry, Jolly was elated. He had long struggled in his dye business with unwanted stains and now immediately recognized the potential of using kerosene as a cleaning agent. It didn't take long for Jolly to capitalize on his observation and launch a business, inventing the term *nettoyage à sec*, or dry cleaning, to describe the novel venture.

"Dry" referred to the absence of water, not the absence of liquid. Essentially, clothes were washed in kerosene instead of water. Kerosene, a mixture of compounds containing chains of twelve to fifteen carbon atoms, was readily available from the fractional distillation of petroleum. Unfortunately, it was highly flammable, as were other petroleum distillates, such as gasoline, which were also used in the early days of dry cleaning. Fires in dry-cleaning establishments were commonplace, and something had to be done. W.J. Stoddard, an Atlanta dry cleaner, tackled the problem by trying different fractions of petroleum. He found that a fraction that distilled at a lower temperature than kerosene, composed of over two hundred compounds having from five to twelve carbon atoms joined together in chains or rings, did not ignite as readily. In 1928, Stoddard solvent made its entry into the dry-cleaning business. But it wasn't ideal—it was still flammable and smelly.

And then along came the chlorinated hydrocarbons, synthetic compounds that were not flammable and were more effective cleaning agents than the petroleum distillates. Carbon tetrachloride and trichloroethylene were the first ones used, but were quickly replaced by the superior perchloroethylene, or perc, which is still the mainstay of the dry-cleaning industry. Clothes are placed into a machine that looks very much like an ordinary washing machine but is filled with perc instead of with water. Actually, it isn't only perc—while this solvent is excellent for removing greasy stains, it is not as good at removing water-soluble stains or insoluble soils. So, small amounts of "co-solvents" such as butoxyethanol and iso-propyl alcohol are added, along with a variety of detergents. These are similar, but not identical, to the detergents used in regular laundering. They are molecules that have distinct water-soluble and perc-soluble regions and help remove stains that would not normally be soluble in perc by forming a link between the stain and the solvent. In general, one end of a detergent molecule bears a charge and is attracted to insoluble soils, the particles of which then become charged and repel each other. The end result is that dirt is lifted from the surface.

There is no doubt that perc cleans well and is quite kind to fabrics. But it is not so kind to the people who work with it. Various studies have shown an increased incidence of cancer among workers in the dry-cleaning industry, and there have been episodes of escaping solvent sickening people. In one instance in Holland, an accidental release of eighty litres of perc caused fifteen people living in the apartments above a dry-cleaning establishment to be hospitalized with respiratory problems. Given that about 70 per cent of all perc used eventually ends up in the environment, and that there are some 180,000 dry-cleaning shops around the world that use perc, there is legitimate concern about the future of this technology. Even the trace amounts of solvent that may escape from dry-cleaned clothes stored in closets have raised eyebrows among some toxicologists, but the main issue is about the safety of workers in the industry. Modern perc systems minimize, but do not eliminate, the escape of the solvent, so "greener" alternatives are desirable.

Four possibilities have emerged. An improved version of Stoddard solvent, made by isolating a very specific mixture of compounds containing between ten and thirteen carbons from petroleum, is used by some dry cleaners, and a silicone solvent with the foreboding name of decamethylcyclopentasiloxane is also available. This compound is less toxic than perc but some concerns have been raised about its persistence in the environment and potential harm to wildlife. Propylene glycol ethers are another class of solvents that can be used to remove stains, but their environmental consequences are not clear. And then there is liquid carbon dioxide.

The same carbon dioxide that is raising all those questions about climate change may be the answer to dry-cleaning problems. The gas can be compressed to a liquid and used as a solvent in specially designed washing machines. Liquid carbon dioxide is a very good solvent, but like water, it requires the addition of a detergent for maximum cleaning. Until the 1990s, nobody had come up with a detergent molecule that had a carbon dioxide–soluble end. That's when University of North Carolina chemistry professor Joe DeSimone discovered that detergents containing a fluorocarbon grouping on one end worked well in liquid carbon dioxide. And that's why, in the future, we may not have to worry about toxic and environmentally unfriendly dry-cleaning solvents. But carbon dioxide systems may do more than clean clothes. They may clean out wallets too—the systems are expensive. Being green requires an investment of green.

Actually, the greenest answer might be "wet cleaning." The process uses water and novel detergents in specially engineered machines that control temperature and pressure in such a way that even some materials marked as "dry clean only" can be laundered without being damaged.

THE GREENING OF CHEMISTRY

IN 1942, nylon went to war. American paratroopers dropped from the sky with nylon parachutes and hunkered down in nylon tents. Soldiers on leave contrived to seduce European women with gifts of newfangled nylon stockings. Soon, bristles for toothbrushes, strings for tennis racquets, insulation for wires, bearings for machinery, catheters, sutures, umbrellas, underwear, shower curtains and numerous items of clothing hit the market—all of them made of nylon. Consumers were absolutely taken with the miraculous new material. Nobody asked where the raw materials to make nylon came from, or whether nylon production had any impact on the environment. The benefits of DuPont's nylon were clear, and the company's slogan, "Better things for better living . . . through chemistry," struck a chord with the public. Other products of chemistry were duly welcomed. Antibiotics battled disease-causing bacteria, pesticides and synthetic fertilizers increased agricultural yields, preservatives cut food costs, polychlorinated biphenyls improved electrical transformer performance and chlorofluorocarbons introduced a new era in refrigeration and air conditioning. Life was good, and chemistry basked in the limelight.

And then, in 1962, Rachel Carson's book *Silent Spring* cast a pallor over the burgeoning chemical industry. Synthetic pesticides, Carson claimed, may have increased crop yields, but there would be fewer birds flying over those crops. Pesticides—DDT in particular—interfered with the ability of birds to lay healthy eggs, she maintained. And if birds were falling prey to DDT, could humans be far behind? Other revelations further tarnished the image of chemicals. The tragedy of thalidomide and the dioxin contamination of Vietnam through the use of Agent Orange made headlines. Waste oozing from a chemical company landfill caused the entire community of Love Canal, part of

Niagara Falls, New York, to be abandoned. Then came the Union Carbide disaster in Bhopal, India, in which thousands were killed by the accidental release of methyl isocyanate from a pesticide manufacturing plant. Little wonder that, in national surveys, people began to link chemistry with words like "pollution," "nuclear winter," "toxins" and "poisons." Extreme views, to be sure, but not totally dismissible. In the headlong rush for new products and increased profits, pollution control and safety concerns sometimes took a back seat. Environmentalists and regulators began to see red. We had better start seeing some green, they warned the chemical industry. And the industry took heed. The concept of "green chemistry" was born.

Traditionally, chemical manufacturers identified commercial needs and developed products designed to maximize yields while minimizing ingredient and processing costs. Wastes were disposed of in the cheapest possible way with little forethought. Dilution was regarded as the solution to pollution. The industry's mantra was "Do what it takes to make a profit, and fix any problems that arise when forced to do so." General Electric, for example, dumped its PCB waste into the Hudson River for years. Now the company has to spend millions and millions of dollars to dredge the river to abide by new pollution laws. But that was then. Today, industrial chemists, with help from academics, are rethinking the entire way that chemicals are brought to the marketplace. Fouling your nest eventually comes back to haunt you, the industry has learned. Furthermore, preventing problems is cheaper than fixing them, and looking for sustainable alternatives to petroleum-based raw materials, or feedstocks, is likely to reap benefits in the long run.

Green chemistry is as much a philosophy as a set of rules. It involves a cradle-to-grave approach to dealing with chemicals. Raw materials from renewable sources are preferred, and chemical reactions are designed so as to minimize waste by maximizing the incorporation of atoms from reagents into the desired final product ("atom economy" has become a key phrase). Toxicity of all reagents and products are considered. Biodegradation and environmental-persistence issues are taken into account, and energy requirements for all processes are minimized. Sounds like an unquestionably good thing. Sounds like what should have been done all along. But the truth is that much too often it wasn't. At least not until the U.S. passed the Pollution Prevention Act in 1990 and followed it up by establishing the coveted Presidential Green Chemistry Challenge awards.

In 1998, one of these awards went to the husband-and-wife research team of John W. Frost and Karen M. Draths of Michigan State University for the development of new strains of genetically engineered bacteria that allow petroleum-based feedstocks to be replaced by glucose from plants. These bacteria can be used to convert glucose to adipic acid, a key chemical in nylon manufac-ture. So what's the big deal? The usual synthesis of adipic acid starts with benzene, a petrochemical. Besides being nonrenewable, benzene is carcinogenic, and its industrial use is subject to strin-gent and costly regulations. Furthermore, one of the steps involved in converting benzene to adipic acid results in the release of large amounts of nitrous oxide, a greenhouse gas.

Using glucose as the starting material eliminates the need for ben-zene, as well as the step that generates nitrous oxide. Glucose, of course, is a renewable resource, available from starch or cellulose in plants. Developing large-scale processes based on glucose technology is still a challenge, and it is unlikely that the use of benzene in nylon production can be totally eliminated, but a 20 per cent replacement is possible. The production of vanillin, the main component of arti-ficial vanilla flavouring, is another industrial process that may benefit from this "green" technology. Currently, the making of vanillin relies on the use of catechol, a chemical made from benzene. Given that the worldwide demand for vanillin is roughly 15,000 metric tons a year, a lot of catechol is needed. And that means a lot of benzene. And that means a lot of potential problems—problems that could be greatly reduced if catechol were to be produced from glucose instead of from benzene. Thanks to advances in green chemistry, that is a decided possibility.

UTTER NANOSENSE

YOU JUST KNOW that nanotechnology has arrived as a scientific force when cosmetics hype it, Michael Crichton writes a novel warning about it, Prince Charles castigates it and demonstrators shed their clothes to protest it. Indeed, it must have been quite a scene at a nanotechnology conference in Chicago in 2004 when members of THONG (Topless Humans Organized for Natural Genetics) collectively dropped their pants to expose their rears festooned with the phrase "There's Plenty of Room at the Bottom."

The reference was to Nobel Prize–winning physicist Richard Feynman's 1959 talk at the annual meeting of the American Physical Society, which many believe inspired the age of nanotechnology. Feynman mused about the possibilities of inscribing all twenty-four volumes of the *Encyclopaedia Britannica* on the head of a pin, making tiny electrical circuits and even manipulating single atoms. The latter would be the Holy Grail of chemistry. Today, chemists make new molecules by mixing together appropriate reagents, based on known chemical reactions. But imagine if a new molecule could be constructed by adding atoms one at a time, sort of like building with Lego, except on a very small scale—a *nano*scale! Pretty alluring. Except to the protesters who suggest that such technology would also usher in problems, problems that might not be on a nanoscale.

A nanometre is one billionth of a metre. That's pretty small. It would take a thousand particles, each with a diameter of 100 nanometres, to span the width of a human hair. And this is the scale we are talking about when we talk about nanotechnology, the field of science that deals with substances that have at least one dimension measuring less than 100 nanometres. Why a separate area of study? Because on this scale, materials often behave very differently from their larger counterparts. Take a chunk of gold. You can toss it back and forth between your hands and admire its gilded lustre. Now, take that gold and make it into nanoparticles. These behave very differently—depending on the size and shape of the particles, they show a range of colours from spectacular ruby red to a beautiful purple. How do you get down to nano levels? One way was discovered by Richard Smalley, Harold Kroto and Robert Curl Jr., who in 1985 vapourized carbon with a laser and ended up with a Nobel Prize for their efforts.

These researchers weren't really interested in nanotechnology; they had actually set out to investigate the chemistry of carbon-rich stars. But they made an amazing discovery. Vapourizing the carbon yielded particles that seemed to be made up of clusters of sixty carbon atoms. Smalley, playing with paper models, concluded that these clusters represented a novel form of pure carbon, distinct from diamond and graphite, the two established forms of the element. He proposed that the sixty carbon atoms were linked together in the shape of a sphere, like a soccer ball. Eventually, this novel arrangement of carbon atoms came to be known as buckminsterfullerene, after architect Buckminster Fuller, who had designed a number of geodesic domes. A more affectionate term for these C_{60} molecules was "buckyball." Being about one nanometre in diameter, they really were "nano." Methods were soon devised to join carbon atoms together so that they formed nanotubes instead of nanospheres, and the age of nanotechnology was ushered in.

Buckyballs and nanotubes turned out to have some very interesting properties not found in other forms of carbon. Buckyballs, for example, are effective antioxidants, meaning that they can neutralize those rogue species we hear so much about, the nasty free radicals that form as a by-product of inhaling oxygen and are linked with various diseases as well as with aging. That's why buckyballs have been showing up in some cosmetic products, such as Zelens Fullerene C-60 Day Cream. But there is a question whether or not such a day cream may become a nightmare. Some scientists, including Robert Curl, are concerned that the health effects of nanoparticles have not been sufficiently explored and that under certain conditions, such as exposure to light, buckyballs can generate highly reactive "singlet oxygen," which can be damaging to tissues.

Of course, it is also possible that harnessing the antioxidant potential of fullerenes can lead to some effective drugs. Experiments have already shown that grafting certain chemical groupings onto fullerenes makes them water soluble and, at least in animal models, renders them effective against some conditions linked to free radicals, such as Lou Gehrig's or Parkinson's disease. There is also hope that fullerenes may become useful in ferrying medications into the body and delivering them precisely where they are needed. But that's in the future. There is also a now for nanotechnology.

Remember the unsightly white stuff that protected many a lifeguard's nose from the sun's rays? Well, nano-dispersed zinc oxide in which the particles are about 30 nanometres in size offers superior protection and is totally transparent! Tennis balls coated on the inside with nano-clay platelets offer better air retention and more consistent bounce. Composites reinforced with carbon nanotubes make for stronger golf clubs and tennis racquets. Windows coated with nanoparticles of titanium dioxide don't fog up and actually cause dirt to break down. And then there are the

10-nanometre-long "nanowhiskers." These tiny fibres can be made to bond to fabrics and make them wrinkle and dirt resistant. But not activist resistant. In fact, it was "nanopants," sold at a Chicago store, that raised the ire of THONG and prompted another near-naked demonstration. This time, the protesters anointed their anatomy with "Teflon Is Toxic" signs, apparently believing—incorrectly—that this material was the secret to the stain-resistant nanotechnology. Maybe the THONGsters need to fill their nano-brains with some macro-science.

A CLAIM THAT HOLDS NO WATER

WATERMELON SALES SUDDENLY ROSE IN 2008. Why? Because hopeful men were seduced by media reports proclaiming that watermelon may excite more than their taste buds. Texas A&M, a major American university, put out a press release with the heading "Watermelon May Have Viagra-Effect." News organizations around the world picked up the story, tantalizing their readers with such ingenious headlines as "Watermelon, The New Oyster?" and "Watermelon Could Add Bite to Sex Life." Some even came up with irresponsible but attention-grabbing banners like "Watermelon Can Duplicate Viagra Effects."

Was there any evidence to support these far-fetched claims? No. Did researchers at Texas A&M, or indeed anywhere else, carry out studies to demonstrate that watermelon has any such effect? No. So how did the fanciful headlines come about? Texas A&M carries out extensive research on plant breeding, including that of water-melon. One aim of this research is to investigate the possibility of increasing the naturally occurring amount of an amino acid called citrulline in watermelon. This compound is of scientific interest because, once ingested, it can be converted into arginine, an amino acid that plays a role in immune-system activity as well as in the dilation of blood vessels.

A press officer at the university, given the task of publicizing this rather mundane research, asked breeder Dr. Bhimu Patil to suggest some interesting items about watermelon to be included in the press release. Patil then speculated about the fruit increasing levels of arginine in the blood, and mentioned that arginine is the source of nitric oxide, which in turn dilates blood vessels and can increase blood flow. He also pointed out that Viagra works by increasing levels of nitric oxide. Presto, the press officer put two and two together and came up with five, and the press release about watermelon increasing libido was born. Pure mythology.

While there is zero evidence about watermelon having any effect on male performance, the notion of citrulline increasing arginine levels is correct. That was clearly shown in a study published in *Nutrition*, a respected peer-reviewed journal. Twenty-three volunteers were asked to drink either three or six eight-ounce glasses of watermelon juice a day for three weeks and then had their blood tested for arginine. This amount of juice corresponds to more watermelon than anyone could reasonably eat. Compared with subjects who drank no watermelon juice, levels of arginine increased by 12 to 22 per cent. No physiological consequences of this increase were measured in any way. The point of the research was simply to determine the possibility of increasing arginine levels through diet because of the role of this amino acid in improving blood flow.

Arginine first came to the public's attention back in 1992, when its metabolic by-product, nitric oxide, was named Molecule of the Year by the prestigious journal *Science*. Inspired by *Time* magazine's Man of the Year, the editors of *Science* decided in 1989 to bestow an annual award on a molecule that, in their opinion, had made the greatest impact that year. Not quite as esteemed as an Oscar, but impressive nevertheless.

Why was nitric oxide accorded this honour? Because researchers had shown that it had the ability to relax the muscles in blood vessels and increase blood flow. This is exactly what is desired when coronary arteries become narrowed by arteriosclerosis. The dilation of blood vessels also leads to lower blood pressure—another plus. In fact, this effect had been noted indirectly during the First World War, when munitions workers who had the job of filling artillery shells with nitroglycerin were found to have very low blood pressure. That's because nitroglycerin can break down to form nitric oxide. Today, nitroglycerine is commonly used in the treatment of angina. But nobody knew until 1987 that nitric oxide was made in the body from arginine. Interest escalated with the subsequent discovery that this molecule also played a role in producing erections. Its ability to enhance blood flow at the right time was responsible for this effect. As soon as that observation was made, pharmaceutical companies began to explore methods for increasing nitric oxide levels in the blood. The erection sweepstakes were won by Pfizer, which developed a drug that attacked the problem indirectly. Viagra blocks an enzyme that normally breaks down cyclic guanosine monophosphate (cGMP), the substance that is produced in response to nitric oxide stimulation and is actually responsible for blood vessel dilation.

An obvious question that arises is, Why not just take arginine directly as a dietary supplement? Indeed, there is some evidence that arginine taken orally can allow blood vessels to dilate more readily, but at the dosage required, arginine pills have been linked with nausea and gastrointestinal discomfort. Supplying the body with citrulline, which is then converted into arginine, may get around this problem. That's why Texas A&M researchers are interested in breeding watermelon with higher levels of citrulline. But keep in mind that no studies have shown that boosting arginine, either through supplements or through high-citrulline foods, has any significant effect on erectile biology. The only activity that will be stimulated by increasing watermelon intake is urination. A ripe, sweet watermelon can certainly give a boost to the taste buds, but don't expect anything else to be boosted. And as a final note, most of the citrulline in watermelon is found in the rind, not exactly the focus of consumption. That little tidbit escaped the attention of writers who generated silly headlines like "No Wonder Watermelons Grow So Big." But it probably didn't escape the attention of supplement manufacturers with visions of rising profits from the sale of watermelon rind tablets.

SIP IT GINGERLY

IT'S AN OILY, tasteless liquid that finds widespread use as a flame retardant, a lubricant for airplane engines, a hydraulic fluid and an additive that imparts flexibility to plastics. It has another distinction too: tri-ortho-cresyl phosphate (TOCP) has had more songs written about it than any other chemical. True, its name is not mentioned in any of the songs, but the lyrics make no secret of the often-devastating effects of this substance.

There's no point looking for any reference to TOCP in the latest releases on iTunes. But back in 1930, the public was treated to classics like the "Jake Walk Blues" by the Allen Brothers—"I can't eat, I can't talk, drinking mean Jake, Lord, I can't walk," the brothers crooned. Little did they realize that they were describing the toxic effects of a chemical that had been used as an adulterant in Jamaica Ginger Extract, a popular patent medicine at the time.

Colloquially referred to as Jake, this extract was promoted as a treatment for respiratory infections, menstrual disorders and intestinal gas. But it was the introduction of Prohibition in the U.S. in 1919 that really boosted Jake's popularity. The simple reason was that each two-ounce bottle contained roughly 80 per cent alcohol by weight, making the contents equivalent to about four shots of whisky. And since patent medicines had not been restricted by the Prohibition Act, paralyzing your brain with Jake was perfectly legal. But as thousands of imbibers were to discover, drinking the wrong kind of Jake paralyzed more than just the brain. And not only temporarily.

Our fascinating medical saga begins in February 1930, when a man tottered into a clinic in Oklahoma City, complaining of tingling in the legs and loss of coordination below the knees. "Must have strained it when I lifted an automobile," he told Dr. Ephraim Goldfain. That didn't seem very likely to the young physician, but the symptoms didn't mesh with polio or lead poisoning, either, the classic neurological conditions with which he was familiar. When he saw five more patients that same day with virtually identical symptoms, it became obvious that hoisting automobiles was not the problem. As Goldfain was to learn, hoisting bottles of Jamaican ginger extract was. Something in the concoction was responsible for thousands of cases of a neurological problem destined to be referred to as Jake leg, Jake walk, Jakeitus, Jakeralysis or gingerfoot.

One of the patients who staggered into the clinic that first day revealed that he knew of at least sixty-five others who had become similarly incapacitated. It didn't take long for Dr. Goldfain to track down a number of these cases and establish a link between them. They were all men, mostly living on skid row, and all had sought refuge in bottles of Jake. But ginger extract had been around since 1860, so why was this problem only cropping up now? Goldfain was not equipped to delve into the chemistry of this "type of polyneuritis," but the Federal Hygienic Laboratory (which was just being converted to the National Institutes of Health) and the U.S. Prohibition Bureau did have the required expertise. And when the handful of victims in Oklahoma City were joined by thousands of others across the country, it was time to put this expertise to use.

Scientists at the Federal Hygienic Laboratory quickly determined that samples of Jake collected from victims killed rabbits and paralyzed calves. Curiously, though, monkeys and dogs were unaffected. What could have such a bizarre effect? It was left to chemists at the Prohibition Bureau to identify the offender as tri-ortho-cresyl phosphate. But what was this chemical, sold at the time as a plasticizing agent for lacquers and varnishes, doing in a patent medicine? Trying to dupe government agents, that's what.

The U.S. government was well aware of the enticingly high alcohol content of many patent medicines and looked for ways to discourage consumption for purposes of inebriation. The "active" ingredient in these products, almost always derived from some plant material, usually had a bitter taste. To ensure that the taste was foul enough to deter excessive consumption, legislation was introduced requiring a certain weight of solids to be left behind once the alcohol had evaporated. If the weight of the residue met the requirement, the product's taste was generally nasty enough to prevent indulgence.

Needless to say, such laws were unpalatable to the huge underground industry that sought to subvert Prohibition. As far as Jake was concerned, the challenge was to replace some of the ginger extract with a substance that was not unpleasant tasting, and that would not evaporate when heated by the federal agents. Harry Gross and Max Reisman, brothers-in-law with a history of shady dealings, dedicated themselves to finding the ideal substance. They experimented with several oily liquids with high boiling points, including dibutyl phthalate, a common plasticizer, and ethylene glycol, an antifreeze component. Both of these boiled off too readily. But then a chemist suggested tri-ortho-cresyl phosphate, which at the time was marketed by the Celluloid Corporation. Gross and Reisman contacted Celluloid and were told that the chemical was non-toxic, and off they went. Hub Products, the company they established, purchased enough

TOCP to adulterate 640,000 bottles of Jake. And the rest, as they say, is history.

Eventually, the poisoned bottles of Jake were traced to Hub Products, and Gross and Reisman were indicted for violations of the Pure Food and Drug Act and the Prohibition laws. Because of various legal loopholes, Reisman got off with a suspended sentence and Gross served only two years. Hardly a punishment that fit the crime of paralyzing the legs of thousands of victims. And for many, more than legs were affected—they suffered from "limber trouble." What was that? Well, the lyrics of "Jake Liquor Blues" are pretty descriptive: "That man of mine got limber trouble, and his lovin' can't do me any good." Tri-ortho-cresyl phosphate was a real downer.

BRIEFS OF FRESH AIR

HER CLIENT SEEMED UNCOMFORTABLE. "Is something wrong?" asked Colorado psychologist Arlene Weimer. "Well," the client sheepishly muttered, "it sort of smells like a sewer in here." Weimer wasn't shocked by the comment. After all, she had seen others wrinkle their noses during a treatment session before, but this was the first time that anyone had been so blunt about the odour. And it was embarrassing. Why? Because she knew exactly where the disturbing smell was coming from. It was coming from *her.*

Taking the bull by the horns, Weimer explained to her client that she suffered from Crohn's disease as well as from irritable bowel syndrome, and that these conditions conspired to produce gas, which she had to release frequently. Actually, not just one type of gas, but several. Human gaseous emissions are composed mostly of nitrogen, oxygen, hydrogen, methane and carbon dioxide, none of which has a smell. But about one one-thousandth of bowel gas is made up of the potent trio of hydrogen sulphide, methanethiol and dimethylsulphide, each one of which can indeed be reminiscent of sewer gas. Weimer's family had struggled bravely to live with the problem, particularly her husband, who was extensively exposed to the vicious emanations in the confines of the bedroom. But Buck Weimer was technically inclined, and when his wife revealed the embarrassing office encounter, he decided to seek a scientific solution.

Mr. Weimer was aware of the remarkable adsorbent properties of activated carbon. Incidentally, "adsorption" is not a misspelling of "absorption." It is quite a different phenomenon. Adsorption describes the ability of certain substances to attract and then bind chemicals to their surfaces. The larger the surface area, the greater the efficiency of the binding. And when it comes to powdered or granulated activated carbon, the total surface area can be astounding. Just one gram, quite a bit less than a teaspoon, can have a total surface area that would cover more than six tennis courts. How is that possible? Because the activation process riddles each tiny granule with numerous channels to produce a honeycomb-like effect. This means that there are numerous inner surfaces available for binding.

The process of making activated carbon starts by heating a substance such as coconut shells, bamboo, wood chips or corncobs to a high temperature in an airless environment. Cellulose and other carbon-containing compounds are broken down, leaving a residue that is almost pure carbon. This is then activated by exposure to steam, which expands each particle by creating a huge number of internal channels, and thereby an immense surface area. Once activated, the material is ready to be used. And there are uses aplenty. Home water filters remove chlorination by-products as well as other impurities such as lead and mercury by passing the water through an activated-carbon canister. But there's the possibility for uses on a far greater scale. When a recent explosion at a Chinese chemical plant contaminated a river that supplied drinking water to the city of Harbin, authorities not only supplied treatment plants with activated carbon, but also dumped large amounts into the river to adsorb the benzene. One of the methods used to remove caffeine from coffee involves soaking the beans in water until the flavour compounds and caffeine are extracted, then passing the solution through an activated-carbon filter to remove the caffeine. On a smaller scale, deodorizing insoles (Odor Eaters is a popular brand) can be inserted into shoes, and hospital emergency rooms stock activated carbon to deal with some types of poisonings—barbiturates, insecticides, weed killers, nail polish remover and many plant toxins are strongly bound by activated carbon. So are poison gases. The large cylinder on the front of most gas masks is filled with activated carbon.

In fact, it was the presence of activated carbon in gas masks that gave Buck Weimer an idea. Having everyone don gas masks around his wife was clearly not a viable solution, but what about using activated carbon to tackle the problem at its source? With that in mind, Mr. Weimer decided to try the substance where it could do the most good. He designed prototype underwear made of nylon and coated with polyurethane, a material that is impervious to gases. A small triangular exit hole allowed for flatus to escape, but only after passing through a replaceable filter. The material around the waist and legs was elasticized to limit leakage. It took some experimenting, but eventually Weimer came up with a multi-layered filter made of felt, polypropylene, spun glass and activated carbon. The equipment worked so well that Weimer founded a company, Under-Tec Corporation, to produce and market its product, Under-Ease underwear.

Judging by sales, the problem of odorous flatus is widespread. And Under-Ease has been instrumental in helping many people avoid embarrassing situations. Those who shunned places of worship for fear of inopportune gas release are back praying. Children with intestinal problems who were being made fun of in school are now at ease. And as far as Mrs. Weimer is concerned, there is no longer any concern about sewer gas in her office. In fact, she claims she now produces less gas because anxiety about flatulence has been eliminated. And for those who do not want to wear a protection device all the time, there is another option. A Houston company, Ultra-Tech Products, makes a polyurethane cushion that is coated with activated carbon and can eliminate the smell of flatus— only, of course, as long as you're sitting on it.

Buck Weimer has reaped the benefits of his ingenuity. His company is doing well, and the atmosphere in the bedroom of the Weimer household has improved significantly. And to top it off, Weimer received an Ig Nobel Prize, a tongue-in-cheek award that recognizes achievements that first make people laugh and then make them think. The Ig Nobels are intended to celebrate the unusual, honour the imaginative and spur people's interest in science, medicine and technology. Smell-adsorbing underwear certainly fits the bill.

NEW, IMPROVED DETERGENT

WALMART DOES NOT WANT nonylphenol ethoxylates (NPEs) in any of the products it sells. And when Walmart speaks, chemical suppliers listen. NPEs are excellent detergents, and for decades they quietly went about doing their job without garnering much attention. But evidence is mounting that when released into the environment they can break down into compounds that mimic the action of the hormone estrogen. Given an apparent worldwide decline in sperm quality, an increase in hormone-driven cancers and a decline in the age at which young women begin menstruating, any substance that has hormone-like effects becomes a concern. A cleaning product without NPEs would therefore be regarded as safer and "greener." Like many other companies, Walmart is wading into the green waters. Whether this is because of a true environmental consciousness or because of visions of increased profits doesn't really matter. Walmart's stated goal of having its suppliers find substitutes for twenty "chemicals of concern" will benefit the environment and, possibly, human health.

This is not the first time that an ingredient in detergents has run into an environmental problem. Until the Second World War, doing the laundry meant using soap. It cleaned well, as long as conditions were right—water couldn't be too acidic and couldn't contain high levels of magnesium, calcium or iron, the "hardness" minerals. In an acid solution, soaps are converted to fatty acids, which are insoluble and which separate as a greasy scum; in hard water, soap forms an insoluble curd as it reacts with the minerals. The chemical challenge, then, was to develop a substance that cleaned like soap but didn't have soap's pitfalls. Enter the synthetic detergents, or "syndets." Alkylbenzenesulphonates (ABSs), made from petroleum products, were molecules that, like soap, had a water-soluble end and a fat-soluble end. One end would anchor itself in a greasy stain, while the other was attracted to water. Essentially, the detergent formed a connection between the stain and water, allowing oily residues to be lifted from fabrics.

Alkylbenzenesulphonates quickly displaced soap in laundry products, and for about a decade both consumers and the detergent industry revelled in the clever chemistry that had solved a significant cleaning problem. And then the ABS story took a sour turn. Copious suds began to appear in sewage treatment plants as well as in some lakes and rivers. On occasion, people even got a head of foam when they drew a glass of water from their taps. What was going on? It didn't take long for researchers to discover that naturally occurring micro-organisms that were very adept at breaking down soap molecules were stymied by the newfangled petroleum derivatives. The problem was the specific molecular structure of the ABSs. These molecules, like soap, had a long chain of carbon atoms; but unlike soap, they also had some carbon atoms hanging like branches from the main chain. Microbial enzymes couldn't handle these branched-chain ABSs, and they began to build up in the environment—hence the suds. A solution was needed.

If the problem was caused by the "branching," why not design molecules that are linear? This was a relatively simple problem for chemists to solve, and biodegradable linear alkylsulphonates quickly entered the marketplace. Suds vanished from rivers, almost as if by magic! Although the term would not be introduced until decades later, the replacement of alkylbenzenesulphonates with linear alkylsulphonates was indeed an example of green chemistry. After all, green chemistry, as we have previously seen, is nothing more than the development of products and processes that have the least impact on the environment and on human health. Unfortunately, though, such impacts are notoriously tough to predict.

Linear alkylsulphonates solved the problem of sudsing rivers, but they could not be used in all detergent applications. For example, they did not dissolve well in cold water and were therefore not ideal for low-temperature washing. Also, the carrot of a more effective product was being constantly dangled in front of producers' eyes. Could other synthetic detergents be developed that would be more effective at removing oily residues? Dozens and dozens of novel chemicals were formulated, including the nonylphenol ethoxylates (NPEs). These were excellent cleaning agents, could be used to formulate liquid products and worked well in cold water. Furthermore, their use was not limited to laundry products. NPEs became staples in the textile and pulp-and-paper industries, were used as emulsifiers in pesticide production and as degreasers in steel manufacturing. NPEs became one of the darlings of the chemical industry. Nobody at the time could have predicted that these compounds eventually would be castigated as endocrine disruptors. Their molecular structure had no similarity to estrogen, and in any case, the very concept of environmental estrogens had not yet been formulated. Technology to find trace amounts of substances with hormonal effects in the environment was simply not available.

Now that the problem has been identified, it can be solved. Other surfactants can take the place of NPEs, albeit not at the same cost. But these days, consumers recognize the fact that they may have to shell out more greenbacks to be green. Procter & Gamble, for example, has come out with Tide Coldwater, a detergent that contains a novel proprietary surfactant that makes oily soils soluble in cold water. The product also contains newly developed enzymes that can break down proteins and starches at a low temperature. Years of research were needed to develop this technology, but now Procter & Gamble can make the claim that if everyone in New York City washed their laundry in cold water for just one day, the energy saved would be enough to light the Empire State Building for an entire month. That, I would think, is the kind of product that meets Walmart's green requirements. Especially since it is a concentrate that eliminates excess packaging, and is in line with Walmart's directives to its suppliers to reduce packaging. This plan will save 323,000 tons of coal and 67 million gallons of diesel fuel from being burned, and keep some 677,000 tons of carbon dioxide from being produced. Not exactly Kyoto numbers, but a start.

OPRAH'S PICK: POLYSTYRENE

I MUST ADMIT that I've fantasized about appearing on *Oprah*. What an opportunity to demystify chemistry that would be! But I think the closest I'll ever come will be answering a question from one of her researchers. It seems Oprah has an interest in polystyrene, because the researcher wanted to know whether it was true that this plastic will take hundreds of years to break down in the environment. Undoubtedly, the question was motivated by environmental issues, as some municipalities are currently examining the possibility of banning one of polystyrene's major incarnations, those ubiquitous foam food and beverage containers.

This is not the first time that foamed polystyrene has plucked at environmental heartstrings. Back in the late 1980s a grassroots movement caused McDonald's and other fast food restaurants to eliminate foam packaging. At that time, the major concern was over chlorofluorocarbons (CFCs), chemicals used in the production of polystyrene foam, which were being implicated in the destruction of the ozone layer. This is no longer an issue, as CFCs have been replaced by pentane or carbon dioxide, which are more environmentally friendly.

Today, worries about polystyrene focus on the toxicity of chemicals used to make the material, on the possibility of residual styrene leaching into food from containers and on the issue of biodegradability. Certainly, polystyrene is not readily biodegradable. But neither I, nor anyone else, can confirm that it will take hundreds of years for polystyrene to break down in the environment. It has not been around for hundreds of years, so we don't know exactly how long it takes to break down. Based on the evidence we have, however, we can assume that it will indeed take a very long time to biodegrade, especially if the polystyrene ends up in a landfill. But is this an environmental horror? Not at all.

First, a little background information is in order. Polystyrene is one of our most useful plastics and has been commercially manufactured since the 1940s. It is made from styrene, an oily liquid derived from petroleum. This brings up yet another issue about the material: are we squandering some of our non-renewable resources on the making of this plastic? Hardly. Less than a fraction of 1 per cent of our petroleum reserves is needed to produce all polystyrene products. And there are loads of them. When polystyrene is mentioned, most people conjure up images of foam clamshells and coffee cups, but the substance is also widely used to make articles ranging from computer cabinets and toys to insulating panels. Traces of styrene, the raw material used to make polystyrene, are always present in the finished product, and in the case of food and beverage containers, can leach out into the contents, particularly if the food or drink are hot. This is of some concern since laboratory evidence points to styrene as a potential carcinogen. But studies of workers exposed to the chemical have not revealed any increase in cancer rates, so it is most unlikely that drinking that occasional cup of coffee from a polystyrene cup is a problem.

In an ideal world, all our garbage would be chewed up by microbes and converted into harmless substances such as carbon dioxide, simple nitrogen compounds and water. Pile up some leaves, paper and food scraps in the backyard, and you can watch this happen—it's what composting is all about. But throw a foam polystyrene coffee cup on the compost pile and it will remain untouched. Bacteria do not look on it as a tasty morsel—meaning that it is non-biodegradable. Consider, though, that if you dispose of all this garbage in an airless landfill instead of a corner in your backyard, biodegradation is severely impaired. Even "biodegradable" food packaging made of polylactic acid, a darling of environmentalists because it is derived from corn, will not break down. Of course, the situation is different if the garbage ends up in an industrial composting facility, but not much of our waste does.

How big of an issue is non-biodegradability in a landfill? Not very. In fact, it is biodegradation that can lead to substances leaching out of landfills, although modern landfills are constructed to prevent any environmental contamination. Polystyrene products may indeed sit in that landfill for hundreds of years, but they do so harmlessly. And they don't take up much space. Less than 1 per cent of the volume in landfills is occupied by easily compressible polystyrene products. Maybe they deserve the rest after they have helped protect our fragile articles during shipping and provided us with food containers that are hygienic, do not leak and keep hot foods hot and cold ones cold. True, after use, food containers often end up as litter on streets. But that is not the plastic's fault— it is people's fault. Appropriate use and proper disposal is the key. Polystyrene can be readily recycled where facilities exist, and today's food containers can end up as tomorrow's insulating panels that reduce your heating and cooling costs. Unfortunately there are still too many municipalities that are not equipped to recycle polystyrene. While polystyrene does bring up some environmental issues, let's not lose sight of its valuable contributions. Automobile parts, life-saving flotation devices, laboratory equipment, television sets and, of course, all sorts of food containers are made of polystyrene. And what would the people clamouring to rid our world of polystyrene food containers suggest we replace them with? Paper? Bringing home a fruit salad from the deli in a paper container is hardly the solution. Furthermore, paper is no better at degrading in a landfill than polystyrene, and it occupies more space than the plastic. As far as energy investment goes, it takes more to produce paper than polystyrene. So, now I think I'll go and enjoy a yogourt from a polystyrene container. And I'll hang on to that container in case Oprah calls. It can make for a real neat demo. I'll give you a hint: Shrinky Dinks are made of polystyrene.

CHEMISTRY KICKS IN

ONE WOULD HARDLY EXPECT being knocked unconscious by a foreign object to become a cherished memory. Except when that object happens to be a soccer ball rocketing off the boot of

the most famous player in the world. And therein lies a story—
one that begins with the greatest soccer team ever assembled, the
Hungarian national team of the early 1950s. The Magnificent
Magyars led by the incomparable Ferenc Puskás put together a
string of thirty-two consecutive international victories, a feat that
has never been matched. In 1953, they stunned England 6–3 in
Wembley Stadium, the first time England had ever been beaten at
home by a European side. In the rematch in Budapest, England was
embarrassed by a stunning score of 7–1. Little surprise, then, that
just about everyone conceded the 1954 World Cup, to be held in
Switzerland, to the Golden Team.

On the way to Switzerland, the team stopped for a training session in Sopron, the town where I was born. My father somehow
managed to get us into the practice game, and we were actually
allowed to watch from an area beside one of the goals. I don't
remember much about the game, but as you will see, that's understandable. Like everyone else, I was focused on Puskás, whose powerful left foot had beaten opposing international goalies eighty-four
times in eighty-five games. This time, though, he missed the net—
but he didn't miss my head. I remember the ball coming towards
me, and then the next thing I can recall is being helped to a bench,
and then into a taxi. The next day was another memorable one.
My father came home with a present—a soccer ball! Puskás had
sent it, he told me, as a souvenir of the "event." Frankly, I think my
father bought the ball, but I worshipped it nevertheless.

That ball was nothing like the balls being kicked about today.
It was made of leather panels stitched together, with a slit through
which a rubber bladder had been stuffed inside. The bladder was
inflated with a pump, tied up, and the opening laced shut like a
shoe. As I recall, there were a couple of problems with this ball.
When it got wet, it became very heavy from the water that was
absorbed by the leather. Even worse, after a few months of play, it
began to lose its round shape and started to look more like an egg

than a ball. To us, none of this mattered much. After all, we had a ball to play with. And what a ball! One that had (maybe) been touched by the great Puskás!

At the time, of course, I didn't realize that this ball was already the product of a great deal of evolution. Apparently, the ancient Chinese kicked around a leather ball stuffed with animal hair and cork, and sometime in the Middle Ages the British made the first inflatable bouncing ball. Actually, whether it is appropriate to call an inflated pig bladder a ball is debatable, but that is just what was used in medieval kicking games that often involved whole villages. In the meantime, in South America, natives had discovered that the latex oozing out of certain trees could be formulated into small, bouncing balls. These "rubber" balls, though, were sticky and quickly lost their shape.

And then in 1836, along came Charles Goodyear with his rubber vulcanization process. Goodyear discovered that heating latex together with sulphur made the material much less sticky and more resilient. As chemists would later learn, the reason was that sulphur atoms form bridges between the long molecules of natural rubber, allowing the latter to return to their original shape after being stretched. Goodyear made hundreds of rubber products ranging from hats to calling cards, and in 1855 made the world's first vulcanized-rubber soccer ball. The ball, now on display at the National Soccer Hall of Fame in Oneonta, New York, was made of rubber panels glued together at the seams and was used in 1863 for one of the first known organized soccer games in the U.S. A monument to commemorate this epic event stands on the Boston Common, where the game was played.

In England, they were still blowing up pig bladders, but William Gilbert, a boot maker, hit upon the idea of a protective leather covering. And then H.J. Lindon took the progressive step of replacing the pig bladder with one made of Goodyear's vulcanized rubber. Supposedly he was motivated by his wife's death from a lung disease caused by blowing up too many pig bladders. Soon, ball manufacturers found that using leather from the rump of a cow made for a stronger ball, and that interlocking panels that ran in different directions allowed the ball to keep its round shape.

Then, in the 1940s, chemistry kicked in. Researchers at Standard Oil discovered that isobutylene, a substance derived from petroleum, could be polymerized to make a synthetic rubber that went by the name of butyl rubber. This was essentially impermeable to air, putting an end to the frustrating task of having to constantly inflate soccer balls. Butyl rubber also made automatically sealing valves possible, eliminating the need for a laced opening. And then the real revolution came: synthetic leather, made of waterproof polyurethane or polyvinyl chloride, replaced leather, eliminating the problem of balls gaining weight when wet. Layers of cushioning fabrics were soon added between the bladder and the covering, which was now constructed of twenty hexagonal and twelve pentagonal panels stitched together with polyester cord, ensuring perfect roundness. The 2006 World Cup led to yet another innovation. A ball made of only fourteen thermally bonded panels—with virtually no seams—was introduced, one with improved bounce and accuracy. While I appreciate the science behind these balls, they certainly don't have the same emotional appeal for me as the deformed, battered leather ball I had to leave behind when we escaped from Hungary in 1956.

Puskás also found greener pastures in 1956, launching a second spectacular career with Real Madrid in Spain. Unfortunately, he had failed to lead Hungary to the expected World Cup championship in 1954. After decimating the opposition in the preliminary rounds, and leading 2–0 in the final, the Golden Team, with Puskás hobbled by an injury, lost to Germany 3–2. I remember listening to the game on the radio, with the "Puskás ball" at my feet. When Germany scored that third goal, it was like—well, like being hit in the head with a Puskás shot.

That World Cup, as all the others, was played on a grass field. But that may change in the future. At the under-20 World Soccer Championships hosted by Canada in 2007, we witnessed something that we have never seen in the World Cup: artificial turf. The turf looked like grass and, according to the players, almost felt like grass. What a difference from the first synthetic playing surface developed by Monsanto back in the 1960s. ChemGrass (back then it was still acceptable to use a chemical connection in a positive way) was made by melting together nylon pellets and a pigment, then extruding the hot mix through spinnerets to produce ribbons that could be woven into a fabric. It was durable enough, but falling on it was no fun, even though the nylon carpet was supported by a soft layer of polyurethane foam. When it was installed in Houston's Astrodome as AstroTurf, ballplayers had to add "carpet burn" and "turf toe" to their vocabulary. But nobody expressed fears that the turf would launch a chemical attack against the players running all over it.

The possibility of such chemical warfare emerged as a consequence of attempts to improve upon the original artificial turf. In the 1990s, FieldTurf, a Canadian company, came up with a novel approach. Out went the stiff nylon fibres; in came soft, elastic polyethylene fibres lubricated with silicone oil. These were tufted into a rubberized plastic mat, just like a giant shag rug. The tour de force, though, was the "infill" composed of sand and granules of rubber, which kept the fibres upright and provided shock absorbency. Old rubber tires and athletic-shoe soles were frozen and ground up to provide the required pellets—and, inadvertently, the problems.

The problems revolve around chemicals released from the rubber fill material. Various lead, arsenic and cadmium compounds are used in rubber manufacture and can leach out from the granules into the soil and surrounding waterways, potentially causing harm to aquatic organisms. But of greater concern is the release of polyaromatic hydrocarbons (PAHs) such as benzopyrene, which are known carcinogens. These are present in rubber as a consequence of the addition of carbon black as a reinforcing agent for automobile tires.

Carbon black, which of course is the reason that tires are black, is made by treating a petroleum fraction at a high temperature, a process that also results in the formation of PAHs. Some of these compounds can evaporate as the hot sun beats down on an artificial surface, exposing players to potentially carcinogenic vapours. There is also concern that dust from the rubber pellets can trigger allergies and asthma. And if that weren't enough, some studies have shown that the fearful methicillin-resistant staphylococcus A bacteria (MRSA) survive better on polyethylene than on other surfaces, and can cause infections when players suffer turf burns. So while the new surfaces may look like grass, and even feel like grass, they don't necessarily behave like grass.

FIFA, the world's soccer governing organization, is looking into the possibility of allowing World Cup games to be played on artificial turf and was concerned enough to look into the toxicity matter. After scrutinizing the research, officials concluded that, while tiny rubber granules may release polyaromatic hydrocarbons, the larger ones used as fillers in turf are not a significant source of these compounds. But there is another question about the turf: will Beckham be able to bend it like Beckham? I'd love to see that from the sidelines. I would take the risk of being beaned again. I could even compare the effect of polyurethane with that of leather.

THE FINAL CURTAIN?

THE EMAILS AND PHONE CALLS I GET these days are relentless. Usually triggered by the latest scary headline, panicked consumers want to know how to avoid all the chemicals (a term they often use synonymously with toxins) to which we are exposed. Well, I tell them, I worry a lot more about slipping in the shower than about exposure to chemicals released from my polyvinyl chloride (PVC) shower curtain. I do not look to eliminate plastics marked with the recycling identifier numbers one, three, six or seven from my life, and if I had to wear a shower cap, I would not worry about its effect on my health. Neither would I be concerned about my neighbour making wine from elderberries picked from a bush that grew in a cemetery. All right, now for the details. Because it is there, rather than in headlines, that true science is to be found.

First, the shower curtain scare. This comes to us courtesy of a U.S.–based activist organization that has magnanimously adorned itself with the name Center for Health, Environment and Justice (CHEJ). The Food and Drug Administration it's not. CHEJ commissioned a study of shower curtains made from polyvinyl chloride (PVC), a substance it ominously, and unjustifiably, calls "poison plastic." The study made headlines across North America. Why? Because its results revealed that PVC shower curtains can release as many as 108 volatile organic chemicals. And why is that news? I'm not sure. A cup of coffee will release more than a thousand volatiles. Ah, but the argument, as put forward by CHEJ, is that some of the chemicals released by a shower curtain are classified as hazardous air pollutants by the U.S. Environmental Protection Agency. You know what? Exactly the same can be said about compounds such as furfural, styrene and caffeic acid found in coffee. Yet we do not talk about closing coffee shops or about protecting people from second-hand coffee aroma.

Exposure is not equivalent to hazard. In order to demonstrate that chemicals from shower curtains can cause harm, we need evidence of more than their presence in the air. Did CHEJ carry out any tests to show that these chemicals are absorbed into the body to any significant extent? No. Did the organization examine how such chemicals may be distributed in the body if absorbed or how they are metabolized and excreted? No. Absorption, distribution, metabolism and excretion are the pillars upon which toxicology is built. If these factors are not addressed, arguments about risk crumble. Did CHEJ run any controls to determine what chemicals are present in ambient air? No. The fact is that with the sophisticated instrumental techniques available today, a sample of air will reveal the presence of thousands and thousands of compounds, both natural and man-made. These come from car exhaust, combustion processes, perfumes, cleaning agents, paints, flowers, trees, asphalt, food, sewage, sweat, cooking and, yes, shower curtains. You can include humans among the culprits too. Our flatus contains dozens of volatiles, including highly toxic hydrogen sulphide. So, in light of our overall exposure to tens of thousands of compounds, how likely is it that the chemicals released from PVC shower curtains present a disproportionate risk? Not very. The chlorinated compounds evaporating from the water in the shower probably are of greater concern.

The pliability of PVC shower curtains is achieved by the addition of compounds called phthalates, which act as internal lubricants. Some phthalates are controversial because of their potential hormone-like effects, and indeed certain ones have been eliminated from toys that young children can put in their mouths. The CHEJ study found a number of phthalates in shower curtains, which comes as no great surprise. But the study, apparently because of technical difficulties, did not determine if any of these were released into the air from shower curtains. And most people are not in the habit of eating their bathroom paraphernalia.

So what are the 108 chemicals that got all the publicity? Mostly compounds released from the dyes used to print the coloured designs on the curtains. These compounds would therefore also be released from other plastics, such as polyester or ethylene vinyl acetate (EVA), and even organic cotton, that CHEJ is touting as a replacement for PVC. EVA is a very useful plastic, and doesn't need plasticizers to make it soft and pliable. But it can release vinyl acetate into the air, a compound that the Canadian government has labelled as potentially hazardous. Of course, CHEJ is fixated on the evils of PVC and is oblivious to such details. This is not to suggest that there is any risk associated with EVA shower curtains—any vinyl acetate released would be trivial.

And there is yet another point to be made: have you noticed how a shower curtain is drawn in towards the tub when the water is turned on? That's because the streaming water causes a rapid downward movement of the air, reducing the pressure it exerts on the curtain. The greater pressure on the outside now pushes the curtain inward. Why is this relevant? Because any chemicals released from the shower curtain would be quickly sucked downwards, away from the nose and mouth.

Now, on to the shower cap query. Quite understandable from a consumer's point of view. If vapours from one plastic can invade our privacy in the shower, why not from another, like a shower cap? It's certainly possible in theory, but as far as the danger of shower caps goes, as one concerned lady wanted to know, there is none. Usually made of clear polyethylene, with nary a plasticizer or dye molecule in sight.

What about getting rid of "unsafe" plastics in the home? This question was prompted by a confused article in a suburban paper citing a spokesperson from some local "Environmental Advisory Committee" who stated that #1 plastics are made from PET (polyethylene terephthalate), which is physically hazardous to one's health because it leaches a hormone-disrupting chemical called BPA. This is nonsense—PET is not made with BPA. Polycarbonate plastics, labelled as #7, can leach BPA, but the amounts, with the possible exception of heated baby bottles, are trivial. In fact, polycarbonate bottles are more environmentally friendly because they are not disposable. Single-use PET water bottles have been a marketer's dream but an environmentalist's nightmare. What a brilliant idea to convince people to buy something they don't need and create a multibillion-dollar industry! Never mind that the process uses petroleum resources unnecessarily and that most of the bottles end up in landfills. A crime.

And the elderberry wine question? Well, it highlights the atmosphere of fear that permeates our lives. A lady became concerned when she saw her neighbour spill some elderberry wine he had fermented in his backyard. Why? Because the elderberries had been picked from a bush in a cemetery, where the soil may have become contaminated with the chemicals used to embalm corpses. Could these toxins possibly contaminate her vegetable garden? the anxious caller wanted to know. No. I think we can safely lay this fear to rest.

While most of the scary stories about our living in a "chemical soup" are greatly exaggerated, the stress they cause is very real. As a result, people have become so scared of dying that they forget about living. And if worrying about trace amounts of chemicals being released from shower curtains makes headline news, then living today is pretty good, isn't it? So just take a deep breath and relax. And if you are worried about the trillions of molecules of diverse chemicals you just inhaled, go and take a warm, calming shower. Just don't forget to use a mat—slipping in the shower is a real risk. And don't worry about any phantom risk if the mat is made of PVC.

SCENTS AND NONSENSE

ALTHOUGH I APPRECIATE the clever chemistry involved, I'm not particularly pleased about being led around by the nose. If I smell fresh bread in a bakery, I would like the odour to be coming from the bread, not from a device that dispenses fresh-bread aroma. If I buy a new television set, I would like it to be because of the picture quality, not because I've formed an emotional attachment to a fragrance embedded in the plastic housing. And if I buy a detergent, I would like my choice to be based upon how well it cleans, or on how environmentally friendly it is, rather than on any scent it may leave behind on the clothes. But in a society inundated with consumer products, marketers have to devise more and more ingenious ways to capture a customer's attention. And now they are exploring ways to reach our wallets through our nostrils. Consider the problem faced by a hotel in Florida.

Business in the hotel's basement level ice cream shop was floundering. Then the manager heard about ScentAir Technologies, a company that specializes in scent solutions for businesses. And it had a solution: lead potential customers down the stairs by their nose! The proposal was to waft the scent of sugar cookies into the air at the top of the staircase leading to the basement and the scent of a waffle cone at the bottom. It worked. Bloomingdale's department stores also have sniffed success. ScentAir's lilac, baby powder and coconut fragrances have helped attract customers to its intimate apparel, baby clothes and swimsuit departments.

When the chain of On the Run convenience stores switched its coffee-brewing system to one that used airtight carafes instead of open pots, it saw its sales decline. An obvious answer was to pump the fragrance of freshly brewed coffee into the air, and of course ScentAir was ready to provide this particular aroma. In fact, the company has a library of more than a thousand scents ranging from fresh-cut grass and pink bubblegum to dinosaur breath. The scent of fresh-cut grass isn't too difficult to reproduce because it is mostly due to one compound, cis-3-hexen-1-ol. Dinosaur-breath smell, created for the Children's Museum of Indianapolis to liven up a display on dinosaurs, was more of a challenge. Certainly a little imagination was needed here, since humans never coexisted with dinosaurs and were obviously unable to pass down information about the nature of dinosaur halitosis. Although such concoctions are proprietary, it's a good bet that company chemists resorted to compounds such as skatole (found in feces and bad breath), putrescine (found in rotting flesh) and the delights of hydrogen sulphide or methyl mercaptan produced by oral bacteria.

Sony, the giant electronics company, has also been exploring smells. The idea is to boost sales in this highly competitive area by making the potential customer as comfortable as possible while forming emotional links to Sony products. It's all about "scent branding," an emerging field. Fragrance scientists at ScentAir buckled down, took a bit of vanilla extract and a tad of citrus and combined them with a variety of secret scents to come up with a unique Sony signature fragrance. The hope is that people will associate the pleasant smell with Sony products and develop a favourable attachment to them. And if they do make a purchase, customers will also be going home with scented sachets in their shopping bags to reinforce the connection.

Samsung, a competitor, has also waded into the smell game and features the scent of honeydew melon in its flagship Manhattan store. It remains to be seen whether vanilla or honeydew melon will sell more television sets. But we already know that Nike shoes will sell better if the store has a mixed floral scent and that sales of bakery products rise if the aroma of freshly baked bread is dispensed into the air. And if you are really interested in furthering your nasal experience, you can try "Eau de Rolls Royce 1965 Silver Cloud." This was created when customers complained that later-model Rolls Royces didn't have the same smell as the classic cars. Once again, scent chemists went to work and created a smell that supposedly duplicates the scent of the famous Silver Cloud.

Working for a scent company must be a pretty interesting experience. You never know what customers are going to ask for. A museum may want to duplicate the odour of free-range, grain-fed hog manure, while Tokyo Disneyland is after a honey scent to be used in its Pooh's Honey Hunt ride. Universal Studios in Orlando puts in a most unusual request: it needs the pleasantly foul aroma of ogre flatulence to go along with a Shrek display. Ogre flatulence, though, is a cinch compared with the needs of the detergent industry. In this highly competitive business, consumer choice may depend on the fragrance of washed and folded laundry. But detergents are designed to remove dirt from fabrics, much of which is of an oily character. Fragrances are usually oily materials as well, so how do you design a product that eliminates one type of oily substance but leaves behind another? And how do you make sure fragrance molecules survive the drying process? You can search out compounds that have limited water solubility, or try encapsulating scents in microscopic spheres or resort to microemulsions to enhance the "fragrance experience."

But do we really need to load the environment with an array of chemicals that are not really necessary? If you can smell them, you are inhaling them. What about the issue of triggering allergies or respiratory problems? Pleasant smells are not necessarily harmless. Many ecologically minded people claim that all of this forced smelling stinks and is just another form of pollution. The industry retorts that the chemicals it uses are safe and that perhaps the activists just need some Island Breeze smell to calm them down—it is, of course, commercially available.

SLEEPING WITH THE ENEMY

THEY'RE LITTLE VAMPIRES. They come to life at night and gorge themselves on human blood—possibly yours. But you can forget about scaring them off with a crucifix or with garlic. And don't even think about driving a wooden stake through their hearts. Worst of all, they are not the stuff of legends. They are all too real. *Cimex lectularius*—bedbugs! And they are on the attack.

Let's meet the enemy. Bedbugs are small but readily visible to the naked eye—if you know where to look. And therein lies the problem: the little bloodsuckers are just incredible at finding hiding places. But if you do manage to get a glimpse of one on your pillow when you suddenly turn on a light at night, you'll see a wingless, almost transparent insect, a few millimetres long. That is, unless the bug has feasted, in which case it will be tinted reddish brown by the victim's blood.

Bedbugs are aptly named because they often seek refuge in mattresses, which provide an ideal hiding place from which they can scamper out at night to dine on humans. Bites are not painful, and some people never even realize that they've provided a blood meal. Others may notice little red welts the next day. Reaction to a bite can be very variable, ranging from almost no effect to inflamed, extremely itchy skin. It all depends on an individual's immune response to the ingenious chemistry bedbugs use to turn on the blood tap. The insects' saliva contains several anticoagulant proteins that "thin" the blood for easier consumption but that unfortunately can also trigger an immune response. Fortunately, there is no evidence that bedbugs transmit any kind of disease.

Bedbugs do seem to have a preference for who is to be on the menu. There are numerous accounts of people sleeping in the same bed, with one being besieged and the other left to sleep in peace. Why this happens isn't clear. Sometimes the infestation may be only on one side of the bed. More likely, the bugs are attracted to a certain scent. While no specific human odours that may entice bedbugs have been identified, there is evidence that bugs can pick up smells. They themselves are known to release chemicals (pheromones) that help establish romantic encounters and others that warn confreres of impending danger. Indeed, the presence of these chemicals is one way of finding bedbug infestations, a notoriously difficult task. What is needed is an extremely sensitive instrument that can pick up bedbug pheromones. Such as a dog's nose.

Inspector Kody possesses the appropriate equipment. This canine detective was rescued from a dog pound by Michael Goldman, who runs a pest control company in Toronto. The dog nobody wanted has become a celebrity by nosing out bedbug infestations with an astounding 95 per cent accuracy rate. Human pest-management professionals usually will do no better that 35 per cent. Kody comes into a room, sniffs around and sits down wherever there is bedbug activity. And this isn't necessarily the mattress—bedbugs can squeeze into the smallest cracks and hide in floors, headboards, electrical outlets, drawers, clothes and luggage. In fact, it is their astonishing ability to hide almost anywhere that has resulted in bedbugs making a startling comeback after being almost eradicated in the 1950s.

Back then, if you had an insect infestation problem, you sprayed with insecticides, which were regarded as marvellous chemicals capable of putting an end to the insect scourge that had plagued humanity throughout its history. DDT was particularly effective, and in North America, at least, put the bedbug problem to bed. But then environmental and health issues began to cloud the pesticide picture, DDT was banned and other pesticides became more carefully regulated. At the same time, international travel and trade became more popular, making conditions ripe for a bedbug comeback. Stowed away in the nooks and crannies of luggage, hiding in the folds of imported clothes or tucked into cozy cracks in foreign-made furniture, bedbugs, which can go for months without food, headed to America. And they found refuge in our homes and hotels.

Now, don't start thinking rundown homes and flophouses. Bedbug infestation is not a sign of lack of sanitation. A few hitchhiking bugs on a piece of luggage can convert a luxury hotel room into an all-you-can-eat bedbug buffet. And they can send hotel managers scurrying for a pest-control professional. There's no sense trying amateur techniques. Household insecticides are no match for bedbugs, and locating the insects calls for an extremely thorough professional inspection. Mattress seams, bed frames, furniture, baseboards, phone jacks, television sets, picture frames, curtains, carpets and even stuffed animals have to be examined for the bugs themselves or for telltale signs of activity such as fecal spots and cast-off skins. As soon as a hotbed is found, it has to be dealt with before the bugs can escape.

An experienced pest-management professional will start by using a vacuum and a steamer to cut into bedbug populations. Infested articles can also be treated with dry heat or freezing temperatures, but this requires a couple of hours at 45 degrees Celsius or several days at minus-5 degrees Celsius. But none of these treatments provide residual protection, which requires that appropriate insecticides be applied. Depending on the situation, bendiocarb, cyfluthrin, permethrin or various pyrethrins can be used. A follow-up inspection two weeks later is critical to ensure the pests have been eradicated. Covering mattresses and box springs with plastic covers reduces the chance of reinfestation. Clearly, dealing with bedbugs requires a great deal of expertise, and a measure of luck. Now, good night, sleep tight . . . and don't let the bedbugs bite. Ahh . . . easy to say. Unfortunately, if they're there, they will.

AN OVERDOSE
OF NONSENSE

PLACEBO IN THE GRASS

IT MUST HAVE BEEN quite a sight at the World's Columbian Exposition in Chicago in 1893. Clark Stanley, better known as the Rattlesnake King, reached into a sack, plucked out a snake, slit it open and plunged it into boiling water. When the fat rose to the top, he skimmed it off and used it on the spot to create Stanley's Snake Oil, a liniment that was immediately snapped up by the throng that had gathered to watch the spectacle. Little wonder. After all, Stanley had proclaimed that the liniment would cure rheumatism, neuralgia, sciatica, lumbago, sore throat, frostbite and even toothache.

It wasn't too hard to convince the onlookers about the wonders of the liniment, particularly when it came to arthritis. All Stanley had to do was point out that snakes obviously did not suffer from this condition and seemed well lubricated internally. The crowd lapped up the hype and shelled out the money. And many claimed immediate relief from their pain. Could there have been something to this remedy? Maybe. But if it offered any relief, it wasn't because of rattlesnake oil. It seems the snake act was only for show, and the liniment that was actually sold had been previously prepared. And not from snakes. Chemical analysis of a surviving sample revealed a mixture of mineral oil, beef fat, turpentine, camphor and red pepper. As it turns out, both camphor and capsaicin, the latter found in red pepper, do have some painkilling effect when rubbed on aching joints. But the most effective ingredient in Stanley's snake oil was a good dose of placebo.

Actually, Clark Stanley didn't come up with the idea of snake oil as a remedy. That notion can be credited to the ancient Chinese, who rubbed the oil on aching joints and claimed relief. Stanley probably heard about the remedy from Chinese immigrants who had come to America to seek their fortune. Many found jobs building the transcontinental railroad, and could well have used snake oil they had brought along to help deal with the backbreaking work.

Chinese snake oil, though, was certainly not made from rattlesnakes. Traditionally, the oil was extracted from the fat sack of the Erabu sea snake. And that makes things interesting. As it turns out, sea snakes, like fish, are rich in omega-3 fats. Being cold-blooded animals, they have to be equipped with fats that don't harden in cold water, and omega-3 fats fit the bill. Erabu sea snakes are even richer in omega-3 fats than salmon, a popular source of these fats. We've heard a great deal about omega-3 fats in recent years, including their potential benefits in improving brain function, reducing the risk of heart attacks, alleviating depression and even in helping with arthritis.

Ah, the arthritis. There really may be a connection here. Omega-3 fats are the body's precursors to certain prostaglandins that are known to have anti-inflammatory effects. So Chinese snake oil might actually have a beneficial effect. If you ingest it! But rattlesnake oil contains very little omega-3 fat, so even if Stanley's liniment contained some rattlesnake oil, it wouldn't have been much use even if people had swallowed it. But of course, all Stanley asked them to swallow was the hype. This huckster may not have done much for his customers' health, but he did leave us with a legacy. Thanks to him, we commonly use the term "snake oil" for ineffective remedies. And some of today's snake oils make Stanley's product look respectable.

Wild Earth Animal Essences is a case in point. Just picture this scene. Daniel Mapel, a "spiritual psychologist," walks to a clearing in the Virginia woods and places a small bowl of water from a nearby stream on the ground. He then begins to walk in a large circle around the bowl, meditating and invoking the spirit of an animal. I don't quite understand how one attracts animal spirits, but apparently it involves tracing out smaller and smaller circles while mentally asking the animal to share its gifts with humankind. By the time he reaches the centre of the circle, Mapel claims, he and the animal are one. Whatever that means. At this time, he says, the energy of the animal is imparted to the bowl of water. This water is then formulated into essences that are sold as "vibrational remedies" (I kid you not). Each of these, according to Mapel, contains the energetic imprint of the animal from which it is derived. Mapel reassures us that no animals were captured or harmed to develop these products. Phew! I'm relieved to know that no animal has noted the theft of its spirit.

Oh yes, along with the spiritually massaged water, the essences also contain a small amount of brandy as a "vibrational preservative." Without the brandy, we are told, the vibration would quickly dissipate. Maybe without the brandy, Mapel's ideas would also dissipate. But I digress. Lets get down to the essence of the Essence.

A customer can select from a wide array of these wonder products. There's eagle, for "soaring above earthly matters to gain perspective and clarity." But if what you want is "support in creating abundance at all levels of one's life," then you need rabbit essence. If you have a blood pressure issue, then buffalo essence may do the trick, at least according to one testimonial. A patient reports a remarkable drop in blood pressure from this essence, which is recommended "for slowing down and getting in touch with the resonance and rhythms of the Earth." And finally, if you need help "facilitating initiation into the deepest, transpersonal realms of the psyche," then you need snake essence. The meaning of this claim is beyond me, but that is perhaps because I have not availed myself of the recommended five to seven drops of this concoction on a daily basis.

Maybe instead of spiritual vibrations, Mr. Mapel should consider putting some authentic snake oil into his snake essence. Why? Because in two recent studies, dietary sea snake oil enhanced maze-learning ability in mice. In other words, it made the mice smarter. Might be just the remedy our modern snake oil salesman needs to swallow himself.

FITNESS FAD

BERNARD MCFADDEN HAD NO USE for doctors. They couldn't cure any disease, he maintained, but could certainly cause plenty of misery. And he wasn't completely wrong. Back in 1884, when sixteen-year-old Bernard sought treatment for his hacking cough and digestive disorders, the treatments offered were pretty abysmal. Bloodletting was still practised, in spite of the fact that it tended to convert patients into corpses. Purging with mercuric chloride (calomel) was hardly better; it often resulted in a bloody evacuation of the bowels, to say nothing of bleeding gums and mouth sores. By contrast, tobacco-smoke enemas and goose-grease poultices must have seemed positively pleasant. If you got better, it was often in spite of the treatments, not because of them. Little wonder, then, that one frustrated patient referred to physicians as "God's nutcrackers," observing that "they opened the corporeal shell to let the soul escape."

McFadden's frustration with medicine caused him to take matters into his own hands. Exercise, Bernard concluded, was the key to health, and he began to spread the word about "the wild joy that thrilled my nerves when I began to feel that health and strength were surely within my reach." Believing that people would pay more attention to his advice if he spiced up his name, Bernard McFadden became Bernarr Macfadden, guru of "physical culture" and author of the slogan, "Weakness is a crime, don't be a criminal." Macfadden's enthusiasm for exercise and his anti-doctor rants struck a chord with a public hungry for simple solutions to complex problems. His *Physical Culture* magazine was eagerly gobbled up, and by the 1930s Macfadden had built a multimillion-dollar publishing empire.

The exercise maven sure knew how to garner publicity. He sponsored a contest to find the most physically perfect woman in England, then proceeded to marry her. Little did Mary Williamson realize what she was getting herself into. Jumping onto Bernarr's stomach from a high table to demonstrate his fitness in public was just the beginning. She was also expected to buy into some of the more bizarre theories her husband had begun to espouse. "Drugs never cured anything unless you call death a cure," he proclaimed. Diseases were caused by "impurity of the blood," and the cure was fasting. The body, deprived of nutrients, was forced to devour the "impurities." Mary had to endure her children being fasted for whooping cough, measles and chicken pox, and to suffer the ultimate tragedy of watching a son die as Bernarr attempted to cure him of convulsions by plunging the boy into hot water. Other strange ideas followed. Forget the dentist, Macfadden advised, just chew on wood for aching teeth. "Air baths," which amounted to walking around the house naked, were great for health. The male genitals were to be exercised with an instrument of his invention, called the peniscope.

To Macfadden's credit, he decried the use of tobacco and alcohol, taught the superiority of whole-grain products and even started a chain of physical-culture restaurants where vegetarian soups, beans, chopped nuts and vegetable steaks were served. Unfortunately, Macfadden maintained his anti-doctor diatribes into the 1950s, even as medicine began to make significant advances. Still, there is no doubt that his urging the public to engage in exercise reaped benefits that are being confirmed by modern science.

Our immune system, research shows, can be revved up by exercise. People who engage in physical activity catch fewer colds. A study at the University of South Carolina found that among 550 healthy men and women, those who got at least moderate exercise most days averaged one cold a year, while people who were less active reported four colds annually. Even recovery from colds may be hastened by exercise. David Nieman of Appalachian State University studied women with colds and found that those who walked regularly and caught colds recovered in less than five days, while those who were sedentary suffered for seven days. So it seems Macfadden was really onto something with his claim that he never caught a cold because he walked to work twenty miles every day. He must have walked briskly, though, given his theory that one should wear as little clothing as possible, even in winter. Indeed, Bernarr favoured "going commando."

Exercise is great, but more isn't necessarily better. Marathon runners have a greater chance of catching colds for several days after a race. Even with heart health, where the benefits of activity have been clearly demonstrated, it seems that exhaustive amounts of exercise are not needed. Jogging, or walking quickly, for twelve miles a week confers significant cardiovascular benefits. Extending this to twenty miles, while maintaining the same intensity, provides even more benefits. But there is no need to constantly strive for increased speed.

Exercise also plays a role in the fight against that most dreaded of all diseases, cancer. Women with breast cancer who exercise at moderate levels for three to five hours a week have been shown to have higher survival rates than women who exercise less than an hour a week. Studies have also shown that regular physical exercise can slow the progression of prostate cancer. And if you are still not convinced of the value of exercise, consider that it may help keep Alzheimer's disease and other dementias at bay. Researchers at the Karolinska Institute in Sweden checked for the incidence of dementia or Alzheimer's disease in seniors whose exercise habits were monitored for more than thirty years. Those who had engaged in moderate physical activity as they passed through middle age had a 50 per cent lower chance of developing dementia and a 60 per cent smaller risk of Alzheimer's than their sedentary confreres. Macfadden himself was a shining example of the merits of exercise. He earned his pilot's licence at age sixty-three and celebrated his eighty-fourth birthday with a parachute jump. At the age of eighty he married a forty-three-year-old. History does not record whether he needed to make use of his peniscope.

WHAT THEY DON'T TEACH YOU IN LAW SCHOOL

I THINK THAT "NONSENSE" may have become the most frequently used word in my vocabulary in recent years. I can't even begin to guess how often it rolls of my tongue in response to questions I'm asked on my radio show, after public lectures or via email or telephone. Can drinking "oxygenated" beverages increase stamina? Is it true that lemon juice can flush toxins out of the body? Can you cure disease by tuning into its vibrational frequency? What else can you say but "nonsense"? (Well, maybe there is another word . . .) But I must admit I was a little stymied when asked about the claim that Borba Gummi Bear Boosters can help "skin clarity."

Gummi Bears (there are also Gummi Worms) are the rubbery confections made basically of cornstarch, corn syrup, sugar and gelatin. Throw in some food dyes and some citric acid for sour flavour and you've got a child's dream and a parent's nightmare. But what is a Borba Gummi? As it turns out, Borba is not a what but a who. Actually, more than just a who—a whole company named after the who. And who is the who? It is the remarkable Scott-Vincent Borba.

Amazingly, the vast majority of law school graduates do not end up practising law. So what do they do? All sorts of things. Some become writers, while others wend their way into politics. But Scott-Vincent Borba found a unique niche. The Californian has become a "nutraceutical and cosmeceutical entrepreneur," marketing a line of products that purport to improve skin quality both from the outside and the inside. Somehow, I don't think you learn a great deal of chemistry or physiology in law school, but I assume you do learn something about the laws that govern consumer products. And you undoubtedly learn something about the claims you can make on behalf of a product without running afoul of the law. Mr. Borba seems to have learned that lesson well.

Borba now heads a multimillion-dollar company that produces drinks, Gummi Bears and jelly beans that "conspire with your bloodstream to nurture skin where it starts." And of course, these go hand in hand with creams based on Borba's "Fiber-Knit Technology," which incorporates "nutrient-infused micro-hydration and natural hydrolyzed spandex fibres that envelop the epidermis from above and below to create your optimum skin condition." Needless to say, the cream also "revitalizes epidermal firming memory." Of course, as Borba tells us, the cream works best when used in conjunction with his Skin Balance Water.

There is no doubt the man is a marketing genius. While he may not know much about chemical formulas, he sure knows the formula that sells products: take some aging baby boomers who want to retain their youth at virtually any cost, fling some scientific terms like "reverse osmosis," "antioxidants," "calorie-free" and "micro-hydration" at them and offer up some product that contains vitamins, minerals and a smidgen of some natural ingredient that has been in the news. Pomegranate, acai and grape seed work well, but botanicals that nobody has ever heard of are even better. Cherimoya and longan fruit are great, and if you can toss in a few words about how these have been used for millennia by the Chinese, even better. And then claim that you have found the fountain of youth, slap on a hefty price and watch your sales zoom.

How did Borba get into this game? Well, let's let him tell the story. "Honestly, I was rushing out of my house to catch a flight and I tripped with a handful of supplements, a litre bottle of water and a couple skin care products in hand. They all fell into a puddle together and that was my epiphany moment." Just why he was rushing around holding supplements and skin care products isn't clear, but what is clear is that Borba is now selling Skin Balance Waters and Gummi Bears spiked with a blend of antioxidants, vitamins and plant extracts that claim to restore lost beauty. Of course, you must choose the right product, depending on your needs. You will have to drink a different fortified water depending on whether you need "clarifying," "age-defying," "replenishing," "firming" or "skin-calming" effects. I suppose if you have more than one of these problems, there is a risk of becoming waterlogged. And there is another effect. You will have a thinner wallet. The drinks cost $2.50 each.

Is there a chance that the drinks and Gummi Bears can really do what they claim? Not much. No government approval is needed to market such products because they come under the aegis of dietary supplements rather than drugs. That is a strange distinction, because it seems to me that if you make a claim of improving the elasticity of the dermis and removing epidermal toxins, you are making a drug claim. And Borba's Gummi Bear Boosters do claim to "help skin regenerate its natural support system, as they help remove toxins and improve skin clarity with cultivated bio-vitamin complex." This special complex is nothing other than a run-of-the-mill mix of vitamins and inconsequential amounts of grape seed, green tea, chamomile and acai extract. A 375-gram bag of these wonder Gummis is priced at twenty-five dollars, and we are urged not to exceed a bag a day. Well, what can I say about all this? Just one word: nonsense! Actually, I'm not a man of few words. So let me also add *caveat emptor*. I wonder if Mr. Borba learned that in law school.

SENSE AND THE ALTERNATIVES

IT WAS EVIDENT from his glance that the fellow browser in the bookstore recognized me. After a moment of contemplative silence, he lifted his eyes from the book he had been thumbing and addressed me in a friendly fashion. "Why don't you guys at McGill teach this stuff in medical school?"

"And what stuff would that be?" I queried.

"Oh, all this alternative medicine business," came the reply.

"Well," I suggested, "if it were taught, it would no longer be 'alternative.'"

This seemed to confuse him a bit, and probably many of you find it puzzling as well.

"Alternative medicine" is indeed a perplexing term. What does it mean? To me, medicine either works or it doesn't. If it works, it isn't alternative. If it doesn't work, it isn't medicine. So what, then, is alternative medicine? The best definition seems to be "those practices which are not taught in conventional medical schools." And why not? Because medical schools are sticklers for a little detail called evidence. After all, patients have a right to expect that a course of action recommended by a physician has a reasonable chance of working. In science, evidence means statistically significant results from properly controlled experiments, as evaluated by experts in the field. Lack of evidence does not necessarily mean that a particular treatment cannot work, only that it has not been demonstrated to work. And that is when it can be termed alternative. If sufficient proof is mustered, alternative transforms into conventional.

Today, the conventional treatment of ulcers often involves the use of antibiotics. That's because there is now clear-cut evidence that many ulcers are caused by the *Helicobacter pylori* bacterium. When the bacterial connection was first suggested by Drs. Barry Marshall and Robin Warren back in the 1980s, it was certainly in the alternative realm. The conventional wisdom among physicians was that ulcers were caused by stress and excess stomach acid. Skeptics, appropriately, wanted evidence before they jumped on the Marshall–Warren bandwagon. And it didn't take long for it to be provided.

In a cavalier and somewhat foolhardy fashion, Marshall drank a solution of *Helicobacter pylori* and developed a case of gastritis. No ulcer formed, but the experiment managed to stir the scientific community into action, and within a few years hundreds of papers were published on the subject. Controlled trials were carried out, and antibiotics were clearly shown to be an effective treatment for ulcers. Today, this is the preferred treatment and is taught in every conventional medical school. Although some physicians may have scoffed at first at the idea of ulcers being caused by bacteria, they were quickly won over by the evidence. Contrary to what is often claimed by alternative practitioners, physicians are not closed-minded about approaches they may not have learned about in medical school—they just would like to see some sort of evidence of efficacy before advocating them.

Alternative medicine, by the definition I proposed, encompasses a vast array of treatments, ranging from the possibly useful but unproven to the ludicrous. Talking about ludicrous, I eventually picked up the book that the gentleman had been perusing, the one that had stimulated our conversation about alternative medicine. It was written by a chiropractor and alluringly entitled *The Food Allergy Cure: A New Solution to Food Cravings, Obesity, Depression, Headaches, Arthritis and Fatigue.* The thesis of this epic work is that undigested food is the cause of most of our health problems. Given that undigested food does not enter the bloodstream, it is hard to understand how it can be the cause of all ailments. But I guess you have to have a modicum of scientific knowledge to be bothered by this incongruity.

Of course, before allergies can be treated, they have to be diagnosed. The author, Ellen Cutler, describes placing a glass vial of a potential allergen in a patient's hand while pushing down on the opposite, outstretched arm. If the arm is readily pushed down, the patient is allergic. (Plastic vials won't work, as this icon of scientific wisdom points out.) Lest the patient worry about exposure to an allergen, Cutler offers reassurance of safety. The reagents she uses "do not contain the actual substances, but instead are energetic carriers of substance signatures made by various homeopathic suppliers." In simpler terms, her test substances contain nothing. This, though, does not seem to be an impediment in diagnosing allergies.

Once an allergen is identified, a cure can be initiated. As the patient sits holding the vial of offending material, "chiropractic techniques are applied to stimulate the acupuncture site connected to the appropriate meridian." Miraculously (a term that is certainly not in a homeopathic dose in Cutler's vocabulary), the allergies disappear. To anyone with a scientific bent, all of this sounds like a lot of hooey. But that's not why it isn't taught in medical school. After all, Ignaz Semmelweis's suggestion in 1847 that hand-washing could dramatically reduce deaths from childbed fever was also seen as hooey by the medical establishment. But it didn't take long before evidence converted the critics. Nobody today doubts the importance of hand-washing.

The reason that Cutler's allergy diagnosis and treatment are not taught in medical schools is not because there is no physiological basis for the theory, not because she's a chiropractor, and not because of the influence of pharmaceutical companies trying to suppress the information. Quite simply, it is because there is no evidence that they work. The same can be said for treating asthma with hydrogen peroxide, infections with goldenseal or cancer with coffee enemas. Should any of these prove their worth in controlled trials, they will be woven into the conventional curriculum.

If there is no efficacy to these alternative treatments, why do people flock to them? Because alternative practitioners are charismatic and often offer hope where mainstream medicine cannot. They use the placebo effect to great advantage and capitalize on the fact that many diseases are self-limiting and resolve themselves. But when contemplating a course of treatment, it is prudent to reflect upon the words of the late Dr. Victor Herbert, renowned hematologist and champion of evidence-based medicine: "For every complex problem there is a simple solution, and it is always wrong."

DUPED BY CHANDRA

SOMEWHERE IN SWITZERLAND, Dr. Ranjit Chandra is enjoying the good life. With some 120 bank accounts in more than a dozen countries, he obviously has no financial worries. The doctor who for many years managed to deceive the scientific community, the man who conned journalists and consumers alike, the man who stained the name of a university, the man who betrayed science is likely laughing all the way to the bank. And I resent it. Chandra duped me too, and in the process duped the many audiences to whom I related his spectacular findings about vitamin supplements before the fraud came to light.

One of my most often requested public lectures is on the science of aging. Little surprise here, since aging is something we would all rather not do. People are ready to slather themselves with exorbitantly priced creams, inject themselves with hormones, swallow red-wine pills, gulp herbal extracts, drink esoteric juices or just plain starve themselves to preserve their brain function and perhaps squeeze out a few more years of life. Scientific evidence for these schemes is scant, so you can imagine my delight back in 2001 when I came upon a paper published in *Nutrition*, a peer-reviewed journal, entitled "Effect of vitamin and trace-element supplementation on cognitive function in elderly subjects." In it, a Dr. R. K. Chandra of Memorial University in Newfoundland described how he monitored memory, abstract-thinking and problem-solving ability in ninety-six healthy seniors who were given either a placebo or oral supplements of modest amounts of vitamins and trace elements. He found significant improvement with the supplement and recommended that "such a supplement be provided to all elderly subjects because it should significantly improve cognition and thus quality of life and the ability to perform activities of daily living. Such a nutritional approach may delay the onset of Alzheimer's disease." Wow!

Never having heard of Chandra before, I immediately became interested in his research and looked up his previous publications. There were a slew of them—well over a hundred. Chandra had carried out research in areas ranging from infant nutrition to infections in the elderly. I was struck by the breadth of his studies as well as by another unusual feature of his work: a large percentage of his publications were "single author." This is an extreme rarity in a field where studies require the testing and monitoring of large numbers of subjects. But Chandra's curriculum vitae suggested that he was a remarkable man, having received numerous honorary degrees and awards, including the Order of Canada. He had even been funded by the World Health Organization to set up a Centre for Nutrition and Immunology, which became Memorial University's pride and joy. The man was obviously a scientific superstar. When I found that he had also published a paper in 1992 in *The Lancet*, one of the prime medical journals in the world, describing how vitamins and other micronutrients decreased the risk of infection in old age, I was sold. Chandra's work became a feature of my aging lecture, probably prompting some seniors to start taking supplements.

Then one day after one of these presentations I was asked if a supplement called Javaan 50 would be appropriate. I was unfamiliar with this particular product and Googled it. Javaan Corporation, I learned, relied on "published, peer-reviewed medical research to bring nutritional products to market." Its founder was identified as Amrita Chandra. Hmm, I thought. A bit more Googling quickly revealed that Amrita was Dr. Chandra's daughter and that the published literature she referred to was her father's. A little strange, but scientists do set up companies, especially if they have discovered something with public appeal. Interestingly, though, there was virtually no difference between Javaan 50 and other common vitamin-mineral supplements. And then, as I started to look into Dr. Chandra's research, a plethora of disturbing facts began to emerge.

In 2000 he had submitted a paper to the *British Medical Journal* that the referees deemed unbelievable. They asked Chandra to provide the raw data, which he failed to do. The *Journal* then advised Memorial University to look into the matter.

As it turns out, this was not the first time such a request had been made. Back in the late '80s, Chandra had been engaged by several infant-formula manufacturers to study their products. He reported his findings in a publication, much to the surprise of Marilyn Harvey, a nurse who had been engaged to recruit infants for the study. She had been having trouble finding enough babies to get the study rolling, and yet here were the published results!

After some soul-searching, she notified the university. A committee was set up to investigate, but its findings were not published and the university took no action. The fact that Chandra threatened to sue the university for defamation of character played a role.

Despite the *British Medical Journal*'s rejection, Chandra managed to get the rejected paper published in *Nutrition*, which is where I came upon it. *The New York Times* was as impressed as I was and featured an article about Chandra's research. The publicity led experts in the field to have a look at the *Nutrition* paper and conclude that it was fraudulent. Furthermore, they suggested the original study in 1992, on which the data were based, was never done. The university struck another committee and launched another investigation. Meanwhile, Chandra struck back with a reference to an independent study by Amrit Jain, a person nobody could find. His only address was a post office box, curiously near Memorial University. In 2002, Chandra resigned abruptly and moved to Switzerland along with the money he had squirrelled away from sales of his "proven" supplement and grants that had never been used for research. In 2005, *Nutrition* retracted Chandra's "landmark" study, and of course we are left wondering whether any of the studies that brought Chandra fame and fortune had ever been carried out. The emperor, as it turns out, has no clothes. He should have some—striped ones.

FLIGHTS OF FANCY

I MUST ADMIT that I am no expert on quantum mechanics. But I know enough to know that Dr. Deepak Chopra knows even less about the subject than I do. Chopra, you should know, is one of the shining lights of the alternative medicine movement, promising "perfect health" to those who learn to use their mind as a healing force. He actually trained and practised as a traditional physician until he became enchanted by transcendental meditation and the ancient Indian practice of Ayurvedic medicine. He then broke ranks with Western medicine and began to engage in something called "quantum healing."

Although I have read his epic work on the subject, I remain mystified by Chopra's mental machinations. What, for example, are we to make of his response to a question about what exactly quantum healing is? "Quantum healing is healing the bodymind from a quantum level. That means from a level which is not manifest at a sensory level. Our bodies ultimately are fields of information, intelligence and energy. Quantum healing involves a shift in the fields of energy information, so as to bring about a correction in an idea that has gone wrong. So quantum healing involves healing one mode of consciousness, mind, to bring about changes in another mode of consciousness, body."

I really don't know how quantum enters into this parable, but clearly Chopra does not use the term symbolically. He refers to the work of Stephen Hawking and informs us that "once known only to physicists, a quantum is the indivisible unit in which waves may be emitted or absorbed." There is also confusing talk of "tapping into divine consciousness," and then there is the kicker: "by consciously using our awareness, we can influence the way we age biologically . . . you can tell your body not to age." Really? It seems the good doctor may have been miscommunicating with his body. Just take a look at a current photograph of the guru and compare it with one on the cover of *The Quantum Alternative to Growing Old*, published in 1993. It seems to me that considerable aging has been going on.

You've gathered by now that I think Deepak Chopra is full of . . . psychobabble. But some sense can be distilled out of the nonsense he dispenses. No, I don't think you can slow down aging or achieve "perfect health" through quantum gibberish, but yoga and meditation, which Chopra also espouses, can be valuable aids towards well-being. He also makes some sense when he says that "if you have happy thoughts, then you make happy molecules, but if you have sad thoughts, and angry thoughts, and hostile thoughts, then you make those molecules which may depress the immune system and make you more susceptible to disease." Chopra himself, as far as I can tell, has never published any data pertaining to his claims, but others certainly have examined the mind-body connection.

Arthur Stone, an immunologist at the State University of New York, measured immunoglobulin A (IgA) levels in the nasal mucus of seventy-two men. Each day, the subjects were asked to fill out forms evaluating their day, including how often they laughed. IgA is an antibody that marks invading bacteria and viruses for destruction by white blood cells, so higher levels are a sign of a more effective immune system. Stone found that on "happy days," IgA levels were also higher. I guess we could say that there were more "happy molecules." A recent British study also implied that happiness may bring better health. A couple of hundred govern-ment employees were subjected to mental stress tests before hav-ing their saliva analyzed for levels of the stress hormone cortisol and the blood-clotting factor fibrinogen. They were also asked to fill out questionnaires about their degree of happiness. Lo and behold, the least happy people were nearly four times as likely as the happiest to have their fibrinogen increase under stress, and they were also far more likely to see a rise in cortisol levels. Higher fibrinogen is associated with a risk of blood clot formation, while higher cortisol is linked with hypertension.

Then there is the work of Dr. Michael Miller of the University of Maryland Medical Center, who investigated the effect of laughter on circulation. Using ultrasound, he measured the blood flow and extent of dilation in the brachial artery of the upper arm in twenty volunteers as they watched either a serious or a funny movie. When they watched the comedy *Kingpin*, in all but one case the artery relaxed and blood flowed more freely for thirty to forty-five minutes afterward. *Saving Private Ryan*, a decidedly sad and stressful film, had the opposite effect. In fourteen of the volunteers the artery constricted, reducing blood flow. (I guess the six others did not find war that tragic.) The overall results showed that blood flow decreased by 35 per cent after stressful movie clips and increased by 22 per cent after laughter. It seems that laughter can help arteries stay healthy.

Now maybe you see why Deepak Chopra puts me in a quandary. On the one hand, I find his misuse of quantum mechanics irritating, which probably increases my cortisol and fibrinogen levels. On the other hand, he does provide some merriment, which increases IgA levels and improves blood flow. What merriment? Well, it seems Chopra can fly. In his book *Return of the Reishi*, he describes how people who meditate can learn to levitate. He then regales us with his own experience: "I was sitting on a foam rubber pad, using the technique as I had been taught, when suddenly my mind became blank for an instant, and when I opened my eyes, I was 4 feet ahead of where I had been before." Now, *that's* funny. Chopra gets twenty-five thousand dollars for a lecture about such stuff. That's sad. Like I said, Chopra puts me in a quandary.

THE DIRT ON DETOX FOOTBATHS

CHARLATANS ARE CLEVER PEOPLE. They can devise ingenious ways to capitalize on people's fears. And today's worries about environmental pollutants sure do offer up a seductive target. Now, let's be clear: I am in no way suggesting that concerns about pesticides, mercury, lead and environmental estrogens are unwarranted. But nonsensical "detoxifying" gadgets are not the way to address the problem. One of the most reprehensible schemes I've ever come across is the "ionic cleanse" footbath. The victim of this scheme is told that the special electrically powered footbath can remove toxins from the body and improve health. And there is "proof": as the subject sits with his or her feet in the bath, a rust-coloured scum forms, supposedly comprising the accumulated toxins that are being released from the body.

Utter nonsense. What is really happening here is a simple process known as electrolysis. When the power is turned on, a small current flows between a pair of electrodes immersed in the water. In this classic electrolysis experiment, as routinely performed by high school students, oxygen forms at one electrode and hydrogen at the other. But when one of the electrodes is made of iron, as it is in this case, a secondary reaction takes place. The iron is converted to iron hydroxide, which is insoluble in water. Essentially, it is rust. This is what is passed off as the toxins coming out of the body. To enhance the effect, the victim is told to put a small amount of salt into the water to help in the "detoxifying process." This, of course, has nothing to do with detoxifying, but it does increase the conductivity of the water, enhancing the rust formation.

Needless to say, the rusty colour will also be produced when no feet are placed in the water. But the ingenious charlatans can explain this, too: tap water, they say, contains all sorts of toxins that are being removed by the detoxification procedure. To prove this point, they will sometimes demonstrate that distilled water yields no colour. Of course not—distilled water does not conduct electricity. Unfortunately, consumers who are not well versed in science can be very easily taken in by this claptrap. A lady who swam a great deal in chlorinated pools was amazed when she plunked her feet into the footbath and began to smell chlorine. Her interpretation was that this wonder device was removing the toxic chlorine from her body. Not so. When you have salt (sodium chloride) in the water, yet another reaction occurs: chloride gets converted to chlorine, and its scent is readily perceived. But the chlorine is most assuredly is not coming from the person's body.

How is it then that some people feel so much better after undergoing an ion cleanse? Well, that's the placebo effect coming into play. I suspect that if they were told that a special extract of edible wolf peach grown under special hydroponic conditions exposed to photons corresponding to five hundred nanometres had magical curative properties, they would try that as well and sing its praises. But I just described tomato juice.

YOU MUST BE JOSHI-ING

I HAVE NEVER MET DR. NISH JOSHI. In fact, until recently I had never even heard of him. But now, through the magic of the Internet, I feel I know the man intimately. And an interesting man he is. I was put onto his trail by an inquiry about *Dr. Joshi's Holistic Detox*, a book that promises to change one's life forever in just twenty-one days by "avoiding acidic, toxic and refined foods, allowing your body to flush out the toxins as it alters the pH balance from acid to alkaline." This latter claim is scientifically meaningless and immediately raises a red flag. But, as they say, you cannot judge a book by its cover, so a more in-depth examination of Dr. Joshi seemed warranted.

It didn't take long to discover that Joshi was no small potatoes— a food that, incidentally, he advises us to avoid. In fact, in Britain he is quite a celebrity. Dr. Joshi has developed a following based on his reputation as health guru to glitterati such as Princess Diana, Cate Blanchett and Gwyneth Paltrow. But, as I quickly learned, he isn't exactly *Dr.* Joshi. At one time, he was enrolled in medical school in India, but he never completed a degree. Joshi did go on to get a diploma in osteopathy in England, which is quite different from a Doctor of Osteopathy degree in the U.S. American doctors of osteopathy are fully licensed physicians, while British and

Canadian osteopaths are not and do not have prescribing privileges. They deal with musculoskeletal problems and believe that physical manipulation can help with a host of health problems. Osteopaths also dabble in various forms of "alternative medicine" such as home-opathy, reflexology and nutritional "detoxification." The extent of their training is variable, and it seems that Joshi's grasp of nutrition, physiology and chemistry isn't exactly firm.

In spite of a murky medical background, Joshi has remarkable diagnostic techniques, as Rachel Cooke, an investigative reporter for the widely read British newspaper *The Guardian,* discovered. Upon hearing that she felt tired, slept poorly and experienced headaches, Joshi immediately concluded that Rachel had a hormonal imbal-ance, improperly functioning adrenal glands, a chromium deficiency and a "lazy" bowel. All of this without a physical exam or blood tests of any kind! Quite a talented man. And of course he had immediate answers for these problems. First, the reporter's computer had to be "rebooted." Not her laptop—her *inner* computer. A few acupuncture needles stuck into Rachel's temples, wrists and ankles, it seems, were the physiological equivalent of Control-Alt-Delete.

And then came the colonic irrigation. Apparently, Joshi is a believer in the age-old philosophy, first espoused by physicians in ancient Egypt, that "death begins in the colon." The notion is that putrefaction of stools somehow leads to the buildup of poi-sons in the body, a process that can be pre-empted by flushing out the colon regularly. There is no scientific evidence for this, but there is plenty of anecdotal evidence from people who say they feel better after a colonic flush. I suspect that in at least some cases the positive reports are triggered by the hope of avoiding another rectal encounter with a plastic pipe that floods the intes-tines with gallons of water. For Rachel Cooke, the colonic irriga-tion was not a happy experience.

Pumping water through the rear portals on a daily basis is not practical, so other detoxication methods must also be introduced—

such as Joshi's Metabolic Detox pills. The main ingredient here is senna, a plant with powerful laxative properties. Basically, the anthraquinones found in the leaves irritate the bowel and cause spasms that expel the contents. A good thing, according to Joshi, because it means that more toxins (although just what these toxins are is never addressed) are being expelled.

Then came the dietary "detoxication" advice—a curious and somewhat confusing diet in which dairy foods were to be avoided, but yogourt, buffalo mozzarella and cottage cheese were encouraged. Maybe according to Ayurvedic principles, which Joshi also espouses, these somehow do not fall into the dairy category. Sugar and sweets are prohibited, but honey is fine. Which ignores the fact that any nutritional difference between sugar and honey is minimal. Fruits, except for bananas, are to be avoided, along with tomatoes and peppers. Ridiculous. Based on what we know about the benefits of fruit consumption, urging people to give up fruit is nutritional malpractice. Chicken, some fish, tofu, brown rice and dark green vegetables are all right, but red meat and wheat are to be avoided. Aside from people with gluten intolerance, there is no scientific reason to avoid wheat.

In spite of confusing and often silly advice, such as eating carrots, beets, celery and ginger to cleanse the liver, the Joshi Clinic Wellness Centre in London is doing a booming business. An initial consultation with the guru runs four hundred dollars, which doesn't cover any tests—with good reason: there aren't any. But you may come away with a prescription for "cupping." Gwyneth Paltrow did. This ancient Chinese technique involves heating the air inside a cup and placing the inverted cup on the bare skin. Supposedly, disease is caused by the stagnation of chi, the body's inherent energy. Cupping unblocks the flow of chi by creating a vacuum as air inside the cup cools, pulling the skin upward. Paltrow was cupped to "remove toxicity in the body." Although she was pleased with the procedure, reporters had a few toxic comments when the actress showed up at a film premiere in a low-cut gown revealing huge red circles on her back.

To be fair, Joshi's Wellness Clinic also has a gym staffed by knowledgeable trainers, and the meditation classes offered do promote relaxation and stress reduction. But the whole program is intertwined with the sale of nonsensical supplements such as Super Liver Cleanse, Kidney Support and, of course, Metabolic Detox. Joshi does emphasize prevention rather than cure, a noble idea, but the notion that the body can be "detoxified" in twenty-one days is pure folly. So here is my challenge to Joshi: tell me what toxins we are talking about, and let's do a blood test before and after and monitor their presence.

DREAM ON

WE SPEND MILLIONS of tax dollars on MRI imagers, CAT scanners, PET scanners and radiation equipment. Maybe we should rethink the way these funds are spent and use them instead to find a way of cloning Adam McLeod, a young man from British Columbia. Why? Because Adam, it seems, doesn't need any of this sophisticated instrumentation to diagnose or treat illness—he manages to do it just by connecting to a person's "holographic energy system." And the patient doesn't even have to be present! Adam is a whiz at "distance healing," which requires nothing more than viewing a photograph of the subject and remotely adjusting his "quantum hologram." Just imagine the savings to society if we could clone this amazing fellow and have a collection of Adams sitting in a control room, diagnosing and curing people around the world.

But there is an eensy-weensy problem here: so far, the only talent Adam has convincingly demonstrated is the ability to attract crowds. There is no shortage of people willing to plunk down a hundred dollars for a day-long seminar with the man who claims to have cured legendary singer Ronnie Hawkins of pancreatic cancer by treating his tumour on the "energetic level." People flock to his sessions, dreaming of being healed by Adam Dreamhealer, the name McLeod has adopted for his professional career. And quite a career it is turning out to be.

By his own account, Adam first noted his special powers when he was fifteen. All of a sudden, pencils and erasers began to fly out of his hands. Actually, I noted similar effects around the same age. Objects moved strangely around me. Wads of paper mysteriously lifted into the air and metal spoons bent as if they were made of putty. Of course, in my case they were helped by that special something I purchased in magic stores. But I digress. Strangely, no one other than Adam ever saw the pencils and erasers perform their acrobatics, so we have only his word for the occurrence of these kinetic events. I know if I saw objects spontaneously take flight around me, I would make every effort to document the gravity-defying phenomena.

While it seems that Adam alone witnessed the flying school supplies, he did have company when his life was changed by an encounter with a giant black bird that "downloaded all the information in the universe" into his brain. I kid you not—I couldn't make this up. When he was sixteen, Adam had a dream in which a big black bird told him to go to Nootka. He described the dream to his parents, who apparently thought it important enough to research where Nootka was. After all, they were already convinced that their son had special powers, since Adam had healed his mother of trigeminal neuralgia, an excruciatingly painful facial nerve condition. Or so they say. Nootka turned out to be an island near Victoria, British Columbia, accessible only by boat. Off the family went on a wild adventure to search for the big black bird.

Once on the island, Adam knew precisely where to go because the landscape was exactly as in his dream. And suddenly, there it was: a four-foot-tall black bird staring at Adam. Surely a sight that would make any ornithologist's mouth water. In any case, faster than you can say Google, the bird downloaded all the information in the universe, whatever that may mean, into the teenager's brain, and his healing career took flight. If you are interested in more details of this amazing escapade, you can find them in Adam's books. Logic would send you to the Humour section of the bookstore, but chances are you'll find them filed under New Age. Which, for my money, is just an extension of the humour section.

The real question is not how a bird came to accumulate all known wisdom, or why it selected a Canadian teenager to become a human bird brain. The question is whether Adam can really do what he says he can. If so, he will go down in history as the man who rocked the very foundations of science. Let's face it: science just can't explain how looking at a photograph of a person who may be halfway around the world can cause chemical changes in that person's body. What kind of energy can be emitted that doesn't fall off with distance? Since energy cannot be created or destroyed, just changed from one form to another, where is it coming from? And how does it target one specific person? Why isn't everyone in the path of this mysterious force healed?

Adam claims to see "auras" around people. Easy enough to test. All that is needed is a barrier behind which people can hide so that only their auras stick out. Let Adam determine in a statistically significant way whether or not someone is standing behind the barrier. Or gather a number of people with various ailments and have Adam diagnose their illnesses. Simple enough.

The ABC television program *20/20* took a shot at appraising Adam. First, he was asked to evaluate the medical status of a woman volunteer. Adam correctly determined that she had a minor back problem (who doesn't?) but completely missed her breast cancer. He failed to affect a reporter's brain waves, which he claims to be able to do. No documented evidence of unusual healing was turned up by the investigators, and it seems that Ronnie Hawkins's pancreatic cancer was never confirmed by biopsy.

This is not to say that people who flood to Adam's seminars derive no benefit. Belief can be very powerful, and there is no doubt that some people will feel better after one of his healing sessions. But feeling better is not the same as *being* better. That requires clinical evidence. Here is my advice to Adam: show what you can do under proper controlled conditions. James Randi will hand over his million-dollar prize if you can move pencils, detect auras or diagnose illnesses from pictures. A Nobel Prize in medicine, and perhaps even in physics, would surely follow. Otherwise, restrict yourself to motivational speaking and forget about bamboozling people with "quantum holographic healing." That stuff is for the birds.

LAUNDRY BALLS

AS YOU CAN IMAGINE, the ad caught my eye. "Earth Smart Laundry CD works on the principles of quantum physics, not chemistry." I had to have it. So I sent in my sixty dollars and waited. A couple of weeks later, a translucent plastic disc full of a blue liquid—and full of promises—appeared. Never again would I have to use detergents! All I had to do was drop the disc into the washing machine, and through the miracle of "structured water technology," it would "activate the laundry water to mimic the cleaning effect of detergent." There was no mention of what this technology was, or how it involved quantum physics.

I was intrigued. After all, various companies had spent decades working on the chemistry of detergents, concluding that these molecules had to serve a dual purpose. First, they had to alter the surface tension of the water, allowing it to flow more readily and consequently penetrate fabrics with greater ease. Second, the molecules had to form some sort of linkage between water and dirt, allowing the dirt to be rinsed away. This required long molecules, one end of which was water soluble and the other oil soluble. Furthermore, the specific molecular structure was important to ensure that detergents would be biodegradable. Then there was the problem of minerals in the water interfering with the action of the detergent, which required the inclusion of "builders" to tie up these minerals. Phosphates were ideal for this job, but presented environmental problems, triggering a search for replacements. All of this is to say that there is some pretty sophisticated chemistry going on in our washing machines. Could a plastic disc with a blue liquid sealed inside it do the same job as the products that had taken chemists with Ph.D.s decades to develop?

Since the pamphlet that accompanied the disc was not very informative, I decided to call the distributor to ask about the technology involved. After being passed from person to person, eventually I spoke to someone who gurgled something about altering the surface tension of the water. When I queried how a liquid sealed in plastic could do this, and what this had to do with quantum physics, he muttered something about the quantum energy stored in the liquid causing water-molecule clusters to dissociate, allowing small water molecules to penetrate the fabric. I guess "quantum" is baffling enough to have it mean whatever you want it to mean. Silly stuff, of course. Still, as they say, the proof of the pudding is in the eating. So I took a few white T-shirts, rubbed in some backyard dirt, added a few grease stains and headed for the washing machine. No contest. The disc did exactly what I thought it would do: nothing.

That little experiment took place in the 1990s. While it failed, it served at least one purpose. It tuned me in to the existence of such laundry products and stimulated me to collect a few more of these marvels. One type featured magnets and offered the following curious explanation: "When water, or any stream of atoms, enters a powerful magnetic field, it physically changes in the same way atoms change when run through particle accelerators used by physicists. Negatively charged oxygen ions are stripped from stable water molecules and are freed to perform a number of tasks." Gulp! When put to a test, the magnetic discs proved to be unattractive. They performed in the same fashion as the blue liquid–filled ones.

My next encounter was with Geo-Wash, which consisted of three perforated, multicoloured plastic balls with some sort of ceramic squares inside. The "scientific" rationale was that "the kinetic energy of these special ceramics with electrical activity creates a process that assists water to clean clothes." Well, it didn't clean my clothes very well, but it did clean me out of fifty bucks. Next came the Magik Ball, which was said to emit infrared rays that "partition the hydrogenous combinations of water molecules and reinforce the penetration into fabrics." Meaningless mumbo-jumbo. It also "eliminates the water's chloric component." Absurd. Like the Geo-Wash, the Magik Ball contains bits of ceramic, this time in the form of beads that "maintain the pH spectrum at the level of a normal chemical detergent." Incongruous twaddle. But not all the claims are nonsense. I'll buy that the Magik Ball leaves no soap residue and presents no risk of allergy. As far as cleaning goes, it did about as well as plain water. And plain water does a pretty good job, which is why these products can muster testimonials.

Think you've heard enough folly? Well, hang on. The Miracle II Laundry Balls are simply divine. Literally. Clayton Tedeton, an American inventor, claims to have put Godliness into cleanliness. Why Miracle II? Because Miracle I was Tedeton being healed of his injuries suffered in an automobile accident by a televangelist. And apparently God had a reason for healing Clayton Tedeton: he was to become God's cleanliness disciple and save the world from toxic cleaning agents by replacing them with safe, spiritually inspired ones. As Tedeton relates, one day the formulation for these wonder products was miraculously flashed on his bedroom wall.

It seems, though, that heavenly chemistry classes leave something to be desired. Part of the miraculous formula calls for "electrically engineered eloptic energized stabilized oxygenated water." This, along with mysterious ingredients such as "ash of dedecyl solution," is to be used for impregnating the Miracle II Laundry Balls. Apparently these divinely inspired components can also neutralize cobra venom. Not having a cobra in the house, I couldn't test that claim. But as far as cleaning goes, wouldn't you think that God has more important issues to worry about than laundry products? Like the frightening extent of scientific illiteracy? Witness the ABI Laundry Ball. It claims to "manipulate the electric fields associated with hydrogen and oxygen atoms and form crystals in the shape of electrical keys. These keys fit into locks and bonds of other compounds to dissolve away dirt much like the action of enzymes in the human digestive system." Gives me indigestion.

MORE MIRACLE WORKERS

THEY KNOW HOW TO BYPASS A BYPASS. They know how to energize your body and brain without exercise. They can make Parkinson's tremors vanish. They can produce erections that last two hours. And of course, they know how to cure cancer, macular degeneration and diabetes. Who are they? According to the brochures that show up in my mailbox, and I'm sure in yours, they are "maverick physicians," "brilliant MDs" and "courageous doctors who dare to swim against the tide." These health gurus have apparently discovered secrets that thousands of researchers have missed, along with natural remedies that Big Pharma has swept under the carpet. And if you want to reap the benefits of these discoveries, all you have to do is buy their books, subscribe to their newsletters or invest in their dietary supplements.

Dr. Victor Marchione really wants to look after me. That must be why he sends me documents entitled "Health Alert Briefing" or "Confidential Health Briefing," replete with scientific-looking file numbers. And you know the contents must be really important because "action is required immediately." What sort of action? The purchase of Dr. Marchione's Vintality supplement (according to File No. 070 7117) or his Smart Pill (File No. DMF/006756). Obviously, the man must prepare a lot of secret files if such complex numbers are required to keep track of them.

What will Vintality allow us to do? Indulge in food without consequence, energize our bodies and brains without exercise and live healthier without sacrifice. That's right. "Delicious desserts, rich creamy sauces, buttery pastries, and all the other high-caloric and high-fat foods that send your taste buds into a frenzy of joy may no longer be dangerous to your health" as long as you swallow Marchione's pill, which "gives you the health advantages of an entire bottle of red wine in a single tablet!" Of course, you have to swallow his arguments as well. These are mainly based on "exhaustive research at Harvard that nutrients in red wine virtually eliminate the dangers of high-calorie, fat-rich foods." Actually, the research referred to was not done on humans and did not use "red-wine pills."

Harvard molecular biologist David Sinclair fed one group of mice a standard laboratory diet, another group an unhealthy diet with 60 per cent of the calories coming from fat and a third group the same unhealthy diet supplemented with regular doses of resveratrol, an antioxidant found in red wine. As expected, the mice in second group became obese, showed signs of diabetes and heart disease and died prematurely. The mice in the resveratrol group also became fat, but they remained healthy and lived as long as the animals that ate a normal diet and stayed thin. Pretty captivating stuff, but the amount of resveratrol given the mice was roughly equivalent to that found in a hundred bottles of red wine. And it was pure resveratrol, not some ill-defined extract of red wine. Marchione himself admits that until he developed Vintality, you would have to drink hundreds of glasses of red wine every day to give yourself the level of nutrients equal to the amounts used in the research (although he doesn't mention that the research was on mice). Curiously, then, he promotes his tablets, which he openly declares contain nutrients equal to that found in one bottle of red wine. By my count, that equals only four or five glasses, not hundreds.

In any case, we have no idea what these tablets really contain. Resveratrol is notoriously difficult to preserve, and the Vintality label gives no information about specific nutrient contents. And we really can't put too much stock in Marchione buttressing his hype with the impressive longevity of Jeanne Calment, the French lady who set a record by living to the age of 122. According to this astute physician, "scientists now know why cholesterol, high blood pressure, heart problems and immune deficiencies never took their toll on her body. Jeanne Calment loved her red wine and drank it every day!" The good doctor then goes on to tell us that "all the miracles of red wine that kept Jeanne going so strong for so long can be yours even if you never take a sip of the stuff." All you have to do is shell out a hundred dollars for a three-month supply of Vintality. I wonder how many people are going to take Marchione at his word and load up on fatty foods, hoping to neutralize the effects with red-wine pills. Seems to me a pretty dangerous idea for a physician to be promoting.

How does he get such ideas? "I used to work hard to come up with new ideas," he says. "But not anymore. Big ideas come to me more quickly, almost without any effort at all. And it's all because of The Smart Pill. Guaranteed to work for you too." Well, I'm not sure I want to swallow pills that generate ideas about letting people run loose with fat consumption as long as they are taking Vintality supplements. Anyway, what is in these Smart Pills that "can help erase your worries about losing your mental functions as you age or about being stuffed away and forgotten in a lonely nursing home"? Vitamins C (60 milligrams), E (18 International Units), B_6 (5 milligrams), B_{12} (15 micrograms), folic acid (200 micrograms), flaxseed powder (25 milligrams), Siberian ginseng (25 milligrams), ginkgo biloba (15 milligrams), alpha lipoic acid (12.5 milligrams) and spinach leaf powder (12.5 milligrams).

There is absolutely no scientific evidence that this combination has any effect on brain power or memory. The amounts of vitamins are those found in numerous multivitamin supplements, and while there have been some studies attesting to the benefits of ginkgo, ginseng and lipoic acid, they have all used amounts way in excess of that found in the Smart Pill. The only evidence for the Smart Pill is anecdotal. It seems the concoction has allowed Dr. Marchione to come up with a scheme to convince the public to fork out a hundred dollars for a three-month supply of a supplement that is backed by absolutely no scientific evidence. That seems pretty smart.

Dr. Marchione is not the only one interested in my health. Dr. Julian Whitaker, "the world's most acclaimed living founder of natural healing," also sends me his pamphlets. So does Jonathan Wright, a physician "whose brilliant mind finds miracles hidden in substances as harmless as cinnamon, mustard and sugar cane." Then there is Dr. David Williams, who models himself on a fictional character as he "sniffs out bogus health claims and miracle cures like a modern day Sherlock Holmes."

Drs. Whitaker, Wright and Williams are certainly real. And although each one implies that he is the world's leading champion of alternative medicine and its most respected authority, they do have a common belief. They promote the notion that drugs produced by the money-grabbing, unethical pharmaceutical industry are mostly ineffective, cause terrible side effects and drive people into the poorhouse. But natural, highly effective non-prescription remedies devoid of side effects are available. Of course, pharmaceutical companies try to suppress this information, as well as the publications that promote them, in order to protect the sales of their expensive, useless drugs.

Now, I am not so naive as to believe that the pharmaceutical industry is staffed by a bunch of choirboys. There are plenty of skeletons in that closet. Many drugs are hyped beyond what they can deliver, and there are plenty of examples of questionable profit-squeezing. But still, these drugs are based on sound scientific research, and their approval requires the submission of extensive safety and efficacy data. Is the approval process foolproof? Of course not. Some nuances only come to light after a drug has been used in the general population for a long time. No amount of testing can guarantee perfect safety. It is always a question of balancing risks versus benefits. This concept, though, seems to elude the promoters of "natural healing." Their implied message is that if a substance is natural, it is safe. Oh, really? The allergen in peanuts is completely natural. So are poison ivy and the *Amanita muscaria* mushroom. In truth, the safety of any substance can only be determined by observation and testing, not on the basis of whether it was made in a lab or plucked from a bush. And its effectiveness as a drug can only be demonstrated by proper controlled trials.

Julian Whitaker, like the other "natural" prophets, doesn't seem to share this philosophy. Rather, his advice is based on anecdotes, cherry-picked data and preliminary research. "America's #1 Diabetes Doctor," as he is billed in the advertising brochure that promotes his newsletter (I wonder where and when this competition took place), claims to have a "3-day diabetes miracle." I'm not sure what this is, since you have to order his pamphlet to find out, but there is allusion to the use of vanadium and magnesium supplements. While there is some scientific evidence that these minerals may help to control blood sugar in some cases, nobody has ever beaten diabetes in three days, as "living legend Julian Whitaker" claims in huge red letters on the front cover of his "Health and Healing" promotional brochure.

It may take him only three days to cure diabetes, but dementia is tougher. That takes thirty days, according to another of his pamphlets. Of course, you'll need a specific course of vitamins, which he happens to have available for sale. Whitaker also has a vitamin cure for cancer, and can eliminate the need for a bypass operation with fish oil, vegetable juice and B vitamins. I'm sure this is interesting news to the Norwegian researchers who administered either placebo or vitamin B supplements to more than three thousand patients with angiographically demonstrated blockages in their coronary arteries. After thirty-eight months (quite a bit longer than the three-week cure Whitaker claims), there was no difference between the vitamin and placebo groups.

Jonathan Wright sounds more like a deity than a physician. "Instead of aiming a chemical howitzer at health problems, Dr. Wright attacks them with the deft precision of a martial arts master, a mineral here, a vegetable there, and today's most feared diseases collapse at his feet." Wow! One of these wonder vegetables is eggplant. According to his brochure, it can cure cancer. "That's C-U-R-E, not just improve one of today's deadliest and scariest cancers, usually in under three months!" Virtually the entire medical community ignored this natural discovery, but one courageous doctor [Wright] broke the news and has spent his career proving that nobody does it better than Mother Nature." Right. And the tooth fairy leaves coins under pillows.

The cancer Wright refers to is skin cancer, and the research involved the use of a specially prepared eggplant extract. Indeed, the product worked better than a placebo and has potential for the treatment of some skin cancers. But the answer to Wright's question, "Could a true cancer cure with a 100 per cent success rate get covered up?" is no. His "special exposé" is pure nonsense. The study is available in the open scientific literature.

Then there is David Williams. "No other scientist has introduced the world to as many important new health discoveries—or blown the whistle on more fakes and mistakes." And we should really trust him. Why? Because, Williams says, "when some bright, young scientist or inventor hits on something exciting, I'm usually the first person he or she calls." Gee, I thought their inclination would be to publish their results in a scientific journal. But Williams doesn't just take their word for a breakthrough. When he "hears of some promising research he checks it out personally, often hopping on a plane, traipsing through a jungle, or examining patients himself to make sure the results are bona fide." If he's satisfied, he brings samples back home, runs his own tests and even tries it out on himself. Yup, that's just the way that science should be done. And what qualifies Williams as a medical messiah? He's a chiropractor. Apparently a very brainy one. According to his brochure, Williams "connects the dots and ingeniously reveals the big picture, like Albert Einstein." "You could say that he is the Michael Jordan of alternative medicine," the brochure proclaims. You *could* say that, I suppose, but I wouldn't.

A
SENSE OF
HISTORY

GUAYULE BOUNCES BACK

"PERHAPS WE SHOULD REDUCE the average length of a condom by a half," suggested Leon Henderson, head of the Office of Price Administration at a meeting of government officials in 1941. The OPA had been established to prevent wartime inflation and to ration scarce consumer goods such as rubber. Henderson's recommendation was whimsical, but it was seminal to highlighting the extreme effort America would have to undertake to ensure the availability of a sufficient rubber supply for the war effort. Without rubber, the war could simply not be won. A single military plane required half a ton of rubber, each tank needed a ton for military use, and a battleship a stunning seventy-five tons. Trucks and jeeps ran on rubber tires and soldiers needed rubber clothing and footwear.

Even before the war, the U.S. automobile industry had made the country a rubber glutton, consuming roughly two-thirds of the world output. Every day, American cars left some five thousand tons of rubber on the nation's highways, but at the time this aroused no great concern. After all, the supply of rubber from the latex of

Asia's rubber trees was plentiful. That, though, changed dramatically after the attack on Pearl Harbor, soon followed by the Japanese invasion of Malaya, the prime producer of rubber. As American supplies dwindled, the sale of tires and inner tubes to civilians, except physicians, was banned. Speed limits were reduced to thirty-five miles per hour, and the public was urged to donate rubber items, such as bath mats, for recycling. This, however, was just a drop in the rubber bucket. A national crisis loomed, and something had to be done.

There were two possible ways to increase the supply of rubber: find alternate natural sources or develop an effective technology for producing synthetic rubber. As far as alternate sources were concerned, only Liberia and Brazil were of any importance. President Roosevelt's secret trip to Liberia was aimed at establishing a U.S. military base to serve as a springboard for transporting men and military equipment to North Africa—and, just as importantly, to ensure a continued flow of Liberian rubber to the U.S. Roosevelt also struck a deal with Brazilian dictator Getúlio Vargas to provide the U.S. with rubber in return for a host of financial benefits. That agreement turned out to have an ugly side, as thousands of poor Brazilians were duped into signing up as rubber workers in the Amazon with promises of high salaries that never materialized. They worked as virtual slaves, with many dying from tropical diseases and snakebites.

In the U.S. itself, attention focused on guayule, a shrub native to Mexico and the southern states that, like the rubber tree, produces latex. Rubber from guayule was not a novel idea. Long before Columbus came to America, children in what is now Mexico were already playing with balls made from guayule rubber. Commercial rubber production from guayule had even been tried in California in the 1930s, but the project was abandoned because the final product could not compete in price with Malayan rubber. With the war under way, price was no longer an issue, and in 1942 the

U.S. Department of Agriculture launched the Emergency Rubber Project, planting thirteen thousand hectares of guayule bushes in California. It was a massive effort, involving more than a thousand scientists and nine thousand labourers. Although significant amounts of rubber were produced, the importance of guayule faded by 1944 due to the parallel development of synthetic rubber made from petroleum.

The idea of making rubber synthetically dates back to 1879, when Gustave Bouchardat heated isoprene with hydrochloric acid to produce a rubbery substance. Previously, Greville Williams had established that natural rubber was actually a polymer of isoprene, so Bouchardat really did make synthetic rubber. The catch, though, was that the isoprene he used was itself made by decomposing rubber. William Tilden made the first true synthetic rubber when he polymerized isoprene made from chemicals he had isolated from turpentine. During the First World War, Germany manufactured a synthetic rubber known as methyl rubber on a large scale, but it was inferior to the natural product and was abandoned at the war's end. But as Hitler geared up for the Second World War, he put the German chemical industry on a fast track to produce a synthetic rubber that would compete with the natural material in quality. When German armour rolled into Czechoslovakia in 1939, it did so on tires made from butadiene-styrene rubber. The Americans also mastered this technology, and by 1944 were meeting their rubber needs through synthetic production.

As a result, the guayule rubber project was shut down, and the facility that had been built to house the workers was converted to a prisoner of war camp which focused on the "denazification" of the inmates. Today, however, guayule is enjoying a renaissance, not because it can compete with other sources of rubber, but because its latex does not contain allergenic proteins. Allergy to the latex of the rubber tree has become a serious problem with the increased use of latex gloves triggered by the AIDS crisis of the 1980s. Since that time, roughly 10 per cent of health workers have experienced latex allergies, and up to 8 per cent of the general population may also be affected. The Yulex Corporation has come to the rescue, and now it markets gloves made of guayule rubber. Guayule-based pillows, sporting goods and even baby bottles are being developed. And rubber is not the only product that the plant produces. It can also yield energy. After the rubber has been extracted, the shrub can be burned in a power plant to produce steam to generate electricity; alternatively, its cellulose content can be fermented to yield ethanol. Unlike corn, guayule does not compete with the food supply. And yes, condoms made of guayule rubber are available. Full size.

WATER BOMB

YOU WOULDN'T THINK that a rusty old barrel is much of an attraction. But the one that sits near the entrance of the U.S. National World War II Museum in New Orleans manages to draw crowds. Why? Because it represents the death knell for Nazi Germany's attempt to make a nuclear weapon. This particular barrel was one of twenty-nine aboard a ferry crossing Lake Tinnsjø in Norway on its way to Germany. The *Hydro* was sunk by Norwegian saboteurs in 1944, and the barrel languished at the bottom of the lake until archaeologists using a specialized underwater vessel brought it to the surface in 2004. The barrel contained heavy water.

Our story begins in December 1938, at the Kaiser Wilhelm Institute in Berlin, with a true "eureka" moment. Chemists Otto Hahn and Fritz Strassmann were following up a piece of intriguing research carried out by Enrico Fermi in Italy. Fermi had bombarded some common elements with neutrons and discovered that they would sometimes be converted into other elements, although these were always close in mass to the originals. But when Hahn and Strassmann subjected uranium to neutron bombardment, they found a result they could hardly believe: the reaction mixture contained some barium, an element with about half the mass of uranium.

"Perhaps you can suggest some fantastic explanation," Hahn excitedly cabled Lise Meitner, the physicist he had worked with for many years before she was forced to flee Germany for Sweden because of her Jewish heritage. Meitner did not disappoint. Perhaps, she suggested, the uranium atom had been "split" apart. And at that moment, the concept of nuclear fission was born. Further experiments by Hahn, Strassmann and Meitner's nephew Otto Frisch not only confirmed the stunning results, but also revealed that such nuclear fission was capable of releasing tremendous amounts of energy. The military implication of the research immediately became obvious on both sides of the Atlantic.

The German uranium project was led by theoretician Werner Heisenberg and army physicist Kurt Diebner. It was Heisenberg's calculation that slower-moving neutrons would cause the uranium nucleus to split more readily. That focused attention on heavy water, a substance capable of slowing down neutrons. Furthermore, slow neutrons could also be captured by the uranium nucleus, which would then be converted into plutonium, another fissionable element.

As virtually every elementary school student knows, a molecule of water is made of two atoms of hydrogen and one of oxygen— good old H_2O. But not every hydrogen atom in the universe is alike. Some are heavier than others because they have an extra neutron in their nuclei. We refer to these as an isotope of hydrogen and use the specific term deuterium (D). When a water molecule has two deuterium atoms instead of two hydrogen atoms, we have heavy water. Since the natural abundance of deuterium is very low, only one in every 4,500 or so water molecules is heavy.

As early as 1933, Gilbert Lewis at the University of California, Berkeley, worked out a method of separating heavy water from ordinary water. His technique relied on the fact that, under the influence of an electric current, a hydrogen-oxygen bond is broken more readily than a deuterium-oxygen bond. So, when a current is passed through a sample of water, H_2O is broken down more readily than D_2O, leaving the sample enriched in D_2O, or heavy water. Starting in 1934, a plant at Rjukan in the Telemark region of Norway had been using the Lewis process to supply scientists interested in researching the physical and biological properties of heavy water. But in 1939, with the new role of heavy water in nuclear fission, the plant took on great military importance. The German army seized the plant, and barrels of heavy water began to be whisked off to Germany. As far as the Allies were concerned, stopping the shipments became a matter of high priority.

The first effort was made in 1943, when twelve hand-picked, specially trained Norwegian commandos were dropped by parachute onto a mountain plateau in Telemark. Their mission was to blow up the heavy water plant at Rjukan. In one of the most spectacular sabotage operations of the Second World War, the Norwegian commandos managed to destroy enough of the heavy water plant to stop operations, but only for a while. Within six months the Germans had rebuilt the plant. But on November 16, 1943, 140 U.S. bombers swooped in and destroyed the Telemark facility.

The Germans then tried to move the remaining heavy water stock to Germany, but once again the Allies managed to sabotage the venture and blew up the ferry on which the drums were being transported, essentially putting an end to the German nuclear effort. It was one of these drums that was salvaged in 2004 and donated by the Norwegian archaeologists to the New Orleans museum.

How close were the Germans to developing an atom bomb? Not very. The supply of heavy water was limited even before the destruction of the Telemark plant. And Hitler did not place emphasis on the nuclear program until it was too late. At first, the war went well and no "wonder weapons" were needed. But when they were, Hitler looked to rockets. He believed that Europe would crumble in face of the V-2 missiles and put rocket research ahead of nuclear research. Some historians have suggested that Heisenberg and his colleagues deliberately dragged their feet because they did not want Hitler to get his hands on the atom bomb. This is not so, as evidenced by conversations secretly recorded by British intelligence at Farm Hall, a mansion in the English countryside where the German scientists were interned after the war. Hahn was also a prisoner there, even though he had never worked on the nuclear energy program. He was devastated when he heard that atom bombs had been dropped on Japan, believing that his discovery of nuclear fission had led to the deaths of hundreds of thousands.

Hahn took little comfort in the fact that he was awarded the 1944 Nobel Prize in chemistry, a prize that really should have been shared with Lise Meitner. If Hahn was the father of nuclear fission, Meitner was its mother.

YOU CAN'T ESCAPE THE TRUTH

HOUDINI IS FAMOUS for his magic and for his incredible escapes, but he deserves to be remembered also for his tireless efforts to foster critical thinking. It was his mother's death in 1913 that launched Houdini on a second career that would intertwine with his magical performances right up to his premature death from peritonitis at age fifty-two. Houdini had been extremely close to his mother and was devastated when he received the telegram informing him of her death. "A shock," he sighed, "from which I think recovery is not possible."

Although he was the son of a rabbi, Harry Houdini had no religious convictions and had never expressed belief in an afterlife. But desperate people do desperate things. A distraught Houdini began to seek out mediums who claimed to be able to contact the dead. Instead of finding solace in the darkness of the seance room, though, Houdini found rampant fraud. Tables moved, trumpets floated in the air and bells mysteriously rang, supposedly signalling the presence of spirits. While many sitters were amazed by these phenomena, Houdini was greatly angered. He knew exactly what was going on, for the simple reason that he had created similar effects on the stage himself. Levitations and apparitions are classic magic effects and are produced by perfectly explicable scientific means. A livid Houdini now saw his beloved art being used to defraud vulnerable people of their money and resolved to unmask the charlatans.

Harry began to give lectures on the fraudulent methods used by mediums and introduced an exposé of psychics into his full evening show. The audience was first treated to magic, then to some amazing escapes, and finally to a seance where the methods commonly used by mediums were exposed. This was a bit of a touchy business, given that Houdini served as president of the American Society of Magicians, an organization that had a cardinal rule to never expose magic tricks. But the greater good of protecting the public from scams warranted an exception, Houdini argued. In any case, he maintained that he was only revealing effects that were used in the total darkness of a seance, and therefore not of much use on the stage.

In 1924, *Scientific American,* the leading science magazine of the time, established a committee to investigate purported paranormal phenomena. By this time, Houdini had established himself as an expert debunker of nonsense and was asked to sit on the committee. A prize of twenty-five hundred dollars was offered to the first person who, under test conditions, could produce an "objective psychic manifestation of physical character." Had it not been for Houdini, Nino Pecoraro, an Italian medium, might have walked off with the prize.

The magician was away on a lecture tour the first time the committee tested Pecoraro, and in spite of his hands and feet being effectively bound, the medium was able to produce the usual seance manifestations. Trumpets sounded, a bell bafflingly flew through the air and dollar bills materialized seemingly out of nowhere. When Houdini heard about these wonders, he cancelled his tour and returned to New York to take part in the Italian's next test. As the world's foremost escape artist, Harry knew that "tying up" could not be left to amateurs. The use of short ropes was critical because long ones inevitably introduced the potential for some slack. Under Houdini's guidance, Pecoraro's hands were placed into gloves that were then sewn to his underwear. His coat sleeves were also sewn to his trousers, and then he was securely tied with numerous short ropes.

After Houdini finished his handiwork, the committee members sat back and waited for the appearance of the spirits. It was a futile wait. There were no physical apparitions, either. No mystifying bell-ringing. And consequently, no prize money awarded! Many other exposés followed, but neither Houdini nor others who followed in his footsteps were successful in preventing people from believing the unbelievable.

That's why psychics like Sylvia Browne can still take advantage of the desperate. Appearing on *The Montel Williams Show,* Sylvia told a woman that her missing grandchild was alive but had been sold into white slavery in Japan. Sadly, as it turned out, the girl had been murdered in Texas. Among her other uncanny predictions have been that American troops would be out of Iraq by July 2004, that Michael Jackson would be sent to jail and Martha Stewart wouldn't, that a black pope would replace John Paul II. And she failed to see 9/11 coming or Michael Jackson's untimely death.

Sylvia has been repeatedly challenged by magician James Randi, a modern-day scourge of charlatanism, to prove her psychic abilities under controlled conditions. Randi is still waiting. And now, I'll make a prediction: Sylvia Browne will never agree to any test because she knows where her talents really lie—in an amazing ability to convince people that she is an authentic psychic in spite of the lack of any evidence.

HORMONE THERAPY

THE PLACE: CHICAGO. The time: the 1930s. Dr. Charles Huggins left the hospital for his usual short walk home. As he would later recount, his wildly pounding heart forced him to sit down several times during the mile-long stroll. Huggins was a young man with no cardiac problems, but on this day he was overcome by a combination of excitement, happiness and nervousness. Why? Because he had just witnessed a remarkable regression of a prostate tumour in a patient injected with estrogen, the female sex hormone. For the first time ever, the progress of cancer had been hindered by the administration of a chemical. Huggins's excitement was understandable, as were his thoughts: "This will benefit man for ever, a thousand years from now people will be taking this treatment of mine." Well, I'm not sure about the thousand years, but Huggins certainly did usher in the hormonal treatment of cancer that is extensively practised today.

Charles Huggins was born in Halifax, Nova Scotia, graduated from Acadia University and then studied medicine at Harvard. After a surgical residency at the University of Michigan, he was offered a position as a urologist at the University of Chicago, which he accepted despite having no experience at all in this area. Within three weeks he had devoured the standard textbook in urology and gone to work. Almost immediately, Huggins became fascinated with the prostate gland and began to wonder why it increased in size as men aged. Since seminal fluid was known to be composed of secretions from the testes and the prostate, Huggins decided to explore the link between these two glands. The appropriate experiment would involve removing the testes and observing the effect of the surgery on the prostate. Obviously, finding human volunteers presented a practical and an ethical challenge, so thoughts turned to animal models. Beside primates, the only other animals known to suffer from prostate enlargement are dogs and perhaps lions. Why "perhaps"? Because volunteers to carry out rectal exams on lions may be even harder to find than men willing to sacrifice their manhood for the sake of science.

In any case, Huggins decided to focus on man's best friend, although I suspect that in this particular case the friendship may have become somewhat strained. The results were clear: when dogs were castrated, their prostates failed to increase in size. The male sex hormones generated by the testes obviously exercised control over the growth of the prostate. An idea occurred to Huggins: if normal cells in the prostate responded to hormones from the testes, perhaps cancer cells did as well. Huggins knew that dogs, like humans, also develop prostate cancer, and decided to test his theory. Remarkably, when the canine cancer patients' testes were removed, the prostate tumours regressed and in some cases disappeared.

These amazing results convinced Huggins to offer the treatment to human patients who were beyond conventional surgical help. Once more, the results were stunning. In many cases, the progress of the cancer was halted and patients resumed normal lives. Well, not quite normal, but better than the alternative. Huggins pondered whether there might be some more attractive way of countering the male-hormone effect. What about those female hormones that had recently been isolated from the urine of pregnant women? So Huggins began offering estrogen injections to some of his prostate cancer patients, who happily agreed to the trial. This was certainly less invasive than castration. And it worked. The results were extraordinary: tumours decreased in size and the cancer failed to spread. Huggins had clearly shown that male and female hormones had antagonistic effects and further-more that cancer cells did not multiply independently of their environment, as had been believed. Their multiplication could be hindered by hormonal intervention.

If prostate cancer responded to hormones, what about breast cancer? Back in 1878, Scottish physician Thomas Beatson had discovered that rabbits no longer produced breast milk if their ovaries were removed, clearly demonstrating that one organ con-trolled the secretions of another. Maybe, Beatson thought, removal of the ovaries could also stop the secretion of whatever substance was causing breast cancer. Indeed, he found that removal of the ovaries in breast cancer patients often resulted in improve-ment of the disease. Amazingly, Beatson had discovered the stim-ulating effect of estrogen even before the hormone itself had been discovered. Aware of this history, Huggins recognized a parallel to his prostate cancer discovery and confirmed that removal of the ovaries and adrenal glands, which are also a source of estrogen, was an effective treatment for some breast cancer patients. To his disap-pointment, administration of male hormones was ineffective. But why did only some women respond to eliminating estrogen?

To understand the mechanism by which estrogen production was stimulated, Huggins needed the help of a chemist. He con-vinced Elwood Jensen, an organic chemist with experience in steroids, to look into the problem. Jensen's idea was to incorpo-rate radioactive tritium into the estrogen molecule, allowing its journey through the body to be monitored after it was injected. The idea turned out to be brilliant, and Jensen soon found that estrogen was bound by certain proteins in breast tissue that were then labelled as "estrogen receptors." Those cancer cells that were equipped with estrogen receptors ceased to multiply when deprived of the hormone. But some types of cancer cells had lost their estrogen receptors, and these were the ones that failed to respond when the ovaries were removed. Jensen and Huggins then went on to develop a test for the presence of the estrogen receptor, allowing physicians to determine which patients would benefit from surgical removal of estrogen-secreting glands. And when estrogen-blocking drugs such as tamoxifen were discovered, the test indicated which patients would benefit, depending on whether they had an estrogen-positive or -negative tumour.

Charles Huggins's discoveries clearly played a pivotal role in the current treatment of both prostate and breast cancer, and the Nobel Prize awarded him in 1966 was well deserved. A plaque above Huggins desk carried his motto: "Discovery is our business." Luckily for us, he was a good businessman.

ROCKET MEN

ON JULY 20, 1969, you could have launched a *huo chien* down most streets in America without hitting a single person. That's because almost everyone was huddled around a television set, waiting for Neil Armstrong to take his small step for man and giant leap for mankind. What's a *huo chien?* It's a Chinese "fire arrow" that was the prototype for every rocket that has ever flown, including the giant 363-foot-tall Saturn V that propelled Armstrong, Michael Collins and Buzz Aldrin on their epic journey to the moon.

Although the origins of the fire arrows are somewhat shrouded in mystery, it is clear that by about AD 1000 the Chinese were experimenting with arrows fitted with bamboo tubes filled with gunpowder. The tube was sealed at the front, and when the gunpowder was ignited, the hot gases released escaped out the back and propelled the arrow towards the enemy. Eventually, in 1678, Newton would explain the basic principle of rocket propulsion through his third law of motion: "For every action there is an equal but opposite reaction." As far as rocketry is concerned, action refers to the hot gases escaping, and the reaction is the rocket moving in the opposite direction. Of course, you don't have to understand Newton's laws to terrorize an enemy with rockets. As early as 1241 the Tatars used rockets against the Poles in the Battle of Liegnitz, and pirates and navies alike eventually used the incendiary capability of rockets to set fire to sailing vessels. By the late 1700s rockets with metal tubes to hold the gunpowder had been developed in India, providing greater thrust and a range of better than a mile. In the Battle of Seringapatam in 1760, British colonial troops were defeated by a barrage of Indian rockets.

But the British learned their lesson, and under Colonel William Congreve, the art of solid-propellant black-powder rockets was to reach a high point in the early nineteenth century. These Congreve metal rockets varied in weight from twenty-five to sixty pounds, could be fitted with an explosive warhead and had a range of up to three miles. They were used against the Americans in the War of 1812, including during an unsuccessful attempt to capture Fort McHenry, which protected Baltimore's harbour. Francis Scott Key witnessed that attack and memorialized the Congreve "rockets' red glare" in the words to "The Star-Spangled Banner." William Congreve dreamed of designing bigger and better rockets, but he certainly did not dream of using them for space flight. Konstantin Tsiolkovsky did. Looking up at the night sky in Moscow in 1873, he began to wonder if man could ever reach out towards the stars. Although mostly self-educated, he correctly suggested that a liquid fuel, when burned with liquid oxygen, would produce a greater velocity of exhaust gases than gunpowder and could conceivably launch a rocket into space. Tsiolkovsky envisioned manned space flight, complete with such details as retrofire

rockets to slow the space ship's re-entry through the Earth's atmosphere. He, however, was a theoretician and did not carry out any
experiments. But Robert Goddard certainly did. The American
physicist had been attracted to the idea of space flight after reading Jules Verne's classic science fiction novel *From the Earth to the
Moon*. Like Tsiolkovsky, Goddard concluded that liquid fuels were
preferable in powering rockets and calculated that liquid hydrogen
and liquid oxygen were the ideal fuel and oxidizer. By 1919,
Goddard had produced a scientific paper, "A Method of Reaching
Extreme Altitudes," and even suggested that a rocket could be
designed to reach the moon. This earned him ridicule from other
scientists as the "moon rocket man" who did not realize that a
rocket engine could not operate in the vacuum of space. Goddard
demonstrated in the laboratory that rocket engines most certainly
would work in a vacuum, but bitten by the criticism, he became
more secretive about his research. On March 16, 1926, Goddard
made history by launching the world's first liquid-fuelled rocket.
The liquid oxygen-kerosene vehicle only rose to a height of 184
feet, but the launch marked the beginning of the Space Age.

Rocketry had also captivated the imagination of Europeans.
Sparked by Romanian physicist Hermann Oberth's 1923 book *The
Rocket into Interplanetary Space*, the German Society for Space Travel
was established in 1927. But when it was taken over by the army in
the 1930s, space travel took a back seat to military applications.
The Third Reich demanded a rocket capable of delivering a heavy
payload to distant targets. Under the leadership of Wernher von
Braun, the Germans developed the terrifying V-2, which used a liquid oxygen-alcohol system and was capable of hitting targets with
a 1,000-pound TNT warhead at a range of two hundred miles.
The V-2 can be accurately described as the world's first "space"
ship, since it left the atmosphere at the top of its trajectory before
plunging down to its target. No one could see it coming, hear it or
stop it. More than four thousand V-2s were launched at targets in

England and Western Europe, killing and maiming thousands. And that does not include the twenty thousand slave labourers who perished in the factories where the rockets were produced.

After the war, von Braun—and about a hundred of his colleagues—ended up in U.S. hands and went to work for the U.S. Army in Hunstville, Alabama. Here von Braun, whose Nazi past was sanitized in order to make him into an American hero, designed the rockets that would eventually launch American astronauts into space. His efforts culminated in 1969, when the giant Saturn V roared into space, its first stage burning a fuel very much like that used by Goddard's original 1926 rocket. The second and third stages burned liquid hydrogen, the ideal fuel as suggested by Goddard forty years earlier. What a visionary this American physicist had been! In his high school valedictory address in 1904, Goddard had voiced the opinion that "it is difficult to say what is impossible, for the dream of yesterday is the hope of today and the reality of tomorrow." Indeed.

WHEN COKE REALLY WAS THE REAL THING

TODAY, you sure wouldn't expect the Pope, the Queen, or the president of the United States to endorse a commercial product. But back in the late 1800s, two popes, Pius X and Leo XIII, along with Queen Victoria and President William McKinley went on record as devoted fans of Vin Mariani. Widely advertised as a "strengthener of the entire system, and renovator of the vital forces," this wine was promoted as a remedy for "brain exhaustion, nervous depression, sleeplessness and voice fatigue." And it shouldn't come as a surprise that famous actresses such as Sarah Bernhardt, noted writers such as H.G. Wells and Jules Verne, and even the wizard of Menlo Park, Thomas Edison, extolled the virtues of this remarkable beverage. After all, it really did have a stimulating ingredient: cocaine.

Europeans were first introduced to the delights of the coca shrub by the Incas early in the sixteenth century. The natives prized the leaves of this plant above gold or silver, and commonly buried a supply with each nobleman, perhaps hoping to spice up the drudgery of eternity. The Spanish conquistadors were quick to get their hands on this natural treasure and dosed it out to the natives they enslaved to make them work harder and longer on less food. Although the Spaniards brought coca leaves back from the New World, Europeans didn't take to chewing the leaves in the Inca tradition. Some Christian ascetics, though, did use coca to help withstand hunger pangs during fasts. And then along came Angelo Mariani, a Corsican entrepreneur with a background in chemistry.

While working in a Paris pharmacy, Mariani became interested in the potential healthful properties of the coca plant and meticulously gathered whatever scientific information was available at the time. He also experimented with growing the plant in greenhouses and tried various preparations himself. Finally, in 1863, Mariani decided that the best way of delivering cocaine's kick without its usual bitterness was to steep the leaves in Bordeaux wine. This turned out to be a lucky combo, because ethanol improves the experience produced by cocaine, probably through the formation of a metabolite known as cocaethylene. No wonder that people found their fatigue chased away and their aches and pains banished, just like the ads claimed. And there were ads aplenty. Indeed, this is where Mariani's real genius lay. In addition to posters and newspaper inserts, he compiled the *Album Mariani,* a collection of celebrity and physician testimonials, each one accompanied by an artist's sketch of the proponent. Eventually, there were fourteen volumes filled with wonderful accounts about the benefits of Vin Mariani. As a result, the beverage developed a huge following and made Angelo Mariani the biggest importer of coca leaves in the world. It also made him the first cocaine millionaire.

People were now drinking Vin Mariani not only as a remedy for ailments but also as a quick pick-me-up. But pure cocaine picked one up even more quickly. When Albert Neiman successfully isolated pure crystalline cocaine from coca leaves in 1860, pharmaceutical companies began to express an interest in the stimulating properties of the drug. But it was only after Alexander Bennett demonstrated its useful effect as a local anesthetic that cocaine began to appear in various pharmacopeias. By the late 1800s Merck, the pharmaceutical giant, had made it readily available to physicians. It was then that Theodor Aschenbrandt, a German army physician, decided to investigate the effect of cocaine on endurance. He had a supply of the drug issued to Bavarian soldiers and, as he described in a paper published in a German medical journal, was very satisfied with the results.

A young Viennese physician suffering from depression and chronic fatigue read Aschenbrandt's account and injected himself with 50 milligrams of cocaine. He—one Sigmund Freud—was duly impressed with the effect. Cocaine, he said, made him cheerful, eased his worries and gave him energy. Further experiments produced even more dramatic effects. In a letter, Freud warned his fiancée, Martha Bernays: "Woe to you my princess when I come. I will kiss you quite red and feed you till you are plump. And if you are forward, you shall see who is stronger, a gentle little girl who doesn't eat enough, or a big wild man who has cocaine in his body." Freud became a cocaine devotee, recommending it to his patients for everything from digestive disorders to morphine and alcohol addiction. He even endorsed commercial cocaine products, becoming one of the first physicians to receive payment for drug promotion. But when a friend developed a feeling that snakes were creeping over his skin, a characteristic sign of cocaine psychosis, Freud lost his taste for the drug.

But the public's taste for cocaine, however, did not wane. In fact, Vin Mariani began to spawn a host of imitations, including French Wine Coca, invented by an American pharmacist, John Pemberton. In addition to coca leaves, Pemberton also added another stimulant, caffeine, in the form of an extract of the African kola nut. The man was also an advertising wizard. In his hands, French Wine Coca became a remedy for nervous trouble, dyspepsia, all chronic and wasting diseases, and was also a "wonderful invigorator of the sexual organs." It was guaranteed to leave consumers satisfied. With the temperance movement gaining traction in the U.S., Pemberton reformulated his beverage, replacing the alcohol with soda water and adding sugar, vanilla and various flavours derived from fruits and spices. Coca-Cola, an "Intellectual Beverage and Temperance Drink," appeared at soda fountains in 1886! And yes, it really did contain traces of cocaine, albeit the amounts were totally inconsequential. Today, the beverage adorned with the most recognizable trademark in the world is made with coca leaves from which the cocaine has been removed. So, no need to worry about cocaine in your Coke. But the ten teaspoons of sugar in every can—well, that's a different story.

SENSE
VERSUS
NONSENSE

CHANGING THE LIGHT BULB

THE TYPEWRITER'S GONE. So is the record player. Photographic film is on its way out. All have been replaced by superior technologies. Could the incandescent light bulb be the next to go? Maybe. Compact fluorescent lights (CFLs) seem ready to take over with a promise of significant energy savings. But— and it seems there is always a "but"—there already are murmurings about hidden costs. Perhaps even to our health.

The origins of fluorescent lighting can be traced back to a most unusual source. In 1676, French astronomer Jean Picard noted a faint glow emanating from his mercury barometer whenever he moved the instrument. Picard duly recorded this curious effect but was unable to explain it. Forty years later, English scientific-instrument maker Francis Hauksbee finally solved the mystery. He showed that when liquid mercury rubs against glass, it produces a static electrical charge that in turn can cause mercury vapour to glow. Indeed, the discharge of energy by electrically excited mercury atoms is the basis of modern fluorescent lighting. But the journey from Hauksbee's experiments to today's fluorescent tubes and bulbs would take a couple of centuries.

If a bit of static electricity could excite a gas, then surely a high-voltage electric current would be even more effective, figured German glassblower Heinrich Geissler. By the mid-nineteenth century, batteries and electrical generators had become available, allowing Geissler to put his notion to the test. In 1858, with the help of physicist Julius Plucker, he managed to produce an eerie glow by passing an electric current through sealed glass tubes containing small amounts of various gases. These "Geissler tubes" really were no more than electrical toys, but they did arouse the curiosity of Daniel MacFarlan Moore, an electrical engineer. Moore began his career working for Thomas Edison, but tinkering with discharge tubes did not please the boss, who of course owed a large part of his fame to the incandescent light bulb. When Edison asked what he had against the light bulb, Moore, perhaps a little thoughtlessly, replied that "it's too small, too hot and too red." Not an answer to Edison's liking. The two inventors parted company, and Moore started up a business manufacturing fluorescent tubes containing carbon dioxide as the light-producing gas. Unfortunately, these tubes were saddled with technical problems, but Moore deserves credit for introducing the first commercially marketed fluorescent lights.

And then came a turning point, one that harks back to Picard's original observation. American electrical engineer P. Cooper Hewitt demonstrated that mercury vapour was superior to other gases for producing light and thereby laid the foundation for all future fluorescent lighting. Hewitt's tubes were energy efficient but produced an uncomfortable blue-green light. Jacques Risler in France solved that problem in 1926 by patenting a coating that, when applied to the insides of fluorescent tubes, absorbed the ultraviolet light produced by electrically excited mercury atoms and re-emitted it as pleasing visible light. A glowing exhibit of fluorescents at the 1939 New York World's Fair primed the public for the large-scale introduction of tube light-ing, and by the 1950s more light was produced in North America by fluorescents than by incandescent bulbs. Compact fluorescent bulbs were developed by General Electric's Ed Hammer in response to the energy crisis provoked by the 1973 Middle East war but did not become popular until the new century ush-ered in the "green" movement.

Compact fluorescents are indeed "green." They require 75 per cent less energy than incandescent bulbs and can last ten times as long. This amounts to roughly a thirty-dollar savings per bulb and, perhaps more importantly, a significant saving in carbon dioxide emissions. If every home in North America replaced just one tungsten bulb with a fluorescent one, the carbon dioxide savings would be equal to taking a million cars off the road. While it is true that the small amount of mercury in these bulbs is a possible source of pollution, it is more than compensated for by the reduced energy requirement. Coal-fired power plants spew mercury into the air, and if less coal is burned, less mercury is released. Many current compact fluorescents contain as little as one mil-ligram of mercury, which is hardly significant.

As far as environmental effects are concerned, CFLs trump tungsten bulbs—that much is certain. But questions have been raised about the possible health consequences of the energy-saving bulbs. First, fluorescents can flicker, and flickering lights can cause migraines and eye strain. There has also been the suggestion, so far unsupported by evidence, that electromagnetic radiation generated by the electronic circuitry built into compact fluorescents may trigger electrosensitivity, symptoms of which reportedly range from general malaise to joint and muscle pain. And finally, the ultraviolet light emitted by compact fluorescents can cause skin sensitivities. With the drive to moderate climate change by eventu-ally replacing all tungsten bulbs with fluorescent ones, these con-cerns need to be addressed.

Britain's Health Protection Agency investigated ultraviolet radiation produced both by single- and double-envelope bulbs. Single-envelope bulbs are the ones made with clearly visible coils, while double-envelope bulbs look like regular light bulbs. The latter were found to emit essentially no ultraviolet light, while single-envelope bulbs did radiate enough UV to cause skin reddening, but only when exposure was continuous, and only when the source was at a distance of less than twenty-five centimetres from the skin. For anyone engaged in prolonged close work—jewellers, for example—double-envelope bulbs would therefore be the better choice.

As far as the general population is concerned, there need not be any concern over UV emitted by CFLs, with possibly a few exceptions: certain diseases, lupus being a prime example, increase the skin's sensitivity to ultraviolet light. UV exposure that otherwise would have no effect can cause rashes and serious skin lesions in people afflicted with lupus. In this instance, compact fluorescents can be a problem, and patient groups have expressed concern about a total phase-out of regular light bulbs. Lupus patients are not the only ones concerned. As noted above, some people believe that the flickering of CFLs can cause migraines and may even provoke symptoms of electrosensitivity, ranging from malaise to joint pain. We'll examine the validity of these allegations next.

CURRENT AFFAIRS

JUST ABOUT THE ONLY FEATURE of compact fluorescent lights that meets with universal agreement is that the devices produce light. (If anyone wants to contest this, I think I could design an experiment to prove it beyond a shadow of a doubt.) But claims that the flickering of these lights can trigger migraines, or that the electromagnetic radiation they emit makes some people sick, present a more complex scenario for investigators. That's because any symptom a person may feel is determined by an intricate balance of physiological and psychological factors. Add to this the concept of biochemical individuality, the well-established fact that the same intervention may result in very different biochemical activity in different people, and we have a tough nut to crack.

Let's start with the allegation that compact fluorescents can trigger migraines. Flashing lights or stroboscopic effects can cause migraines in susceptible people—about that, there is no doubt. Neither is there any doubt that fluorescent lights do flicker. The question is whether the human brain can detect this flickering. With the old-fashioned fluorescent tubes, quite possibly, since they flicker at a rate of sixty times a second, a perceptible frequency. But compact fluorescents use different electronics and flicker more than ten thousand times a second, a rate not detectable by the human eye. It therefore seems most unlikely that such flickering could be the cause of headaches. Compact fluorescents tend to produce a harsher light, with a greater "blue" component than conventional bulbs, which may be a cause of headaches.

But what if these headaches are not caused by the light's intensity or flickering, but rather by the electromagnetic fields generated by the bulbs? This brings up the thorny and controversial issue of electrosensitivity. People afflicted by this condition claim to suffer from a variety of symptoms, which can include weakness, muscle aches, joint pain, dizziness and headaches, as well as memory and concentration problems when exposed to electromagnetic radiation. Cell phones, cell phone towers, power lines, appliances, computers, television sets and compact fluorescent bulbs have all been blamed for triggering symptoms. Some proponents of the existence of this curious phenomenon have even offered a rationale for the effect, usually explaining that our nervous system functions based on electrical impulses and that it is therefore susceptible to the effects of external electromagnetic fields. But we need more than such simple-minded notions to establish the existence of electrosensitivity.

Of two things we can be certain: people who claim to be electrosensitive do suffer, and all of the items listed above do produce electromagnetic fields. But this doesn't prove the suffering is caused by the fields. Such proof requires evidence that can only be provided by proper controlled studies—not by anecdotes or experiments that border on the childish. Inferring that the risk of electromagnetic radiation from a compact fluorescent light can be determined by the degree of static it generates as a radio is brought near it is nonsense. Electromagnetic fields, as generated by any electrical equipment, can interfere with radio signals. Any student who has ever tinkered with electronics knows that.

Also meaningless are experiments that are not properly blinded. Some researchers who are convinced that electromagnetic fields can be harmful suggest that electrosensitivity effects are due to microsurges in electric current as it flows through wiring ("dirty electricity"). They then claim that the problem can be eliminated by the use of a type of "microsurge filter." In one widely publicized study, such filters were installed in a school, and the experimenters concluded that teachers were less frustrated, less tired and less irritable while they were in place. The teachers' general well-being was said to be improved, as determined by questionnaires.

But the placement of the filters was evident to anyone who cared to look, and the type of questions the teachers were asked made it obvious that there was some experiment underway to study health effects. Surely the researchers would not be installing any device to make people sick, so it is not surprising that the teachers reported feeling better. They assumed that the black boxes must be doing some good. And undoubtedly there was discussion among the teachers about the "experiment." I suspect that even a judge at a high school science fair would raise an eyebrow about the methodology employed by the scientists. It is also discomforting to learn that the lead researcher publishes papers co-authored with the gentleman who markets the filter and uses the results of this highly questionable study as a sales pitch.

Of course, the proper way to carry out an investigation of electrosensitivity is with a double-blind protocol. Enlist subjects who feel that they are sensitive to electromagnetic fields and expose them to such fields without the experimenters or the subjects knowing when the electronic equipment that generates the field is turned on. Then just ask the subjects to detect when they "feel" the field. Since electromagnetic fields cannot be seen, such experiments are not difficult to carry out. And they *have* been carried out, at least thirty-two times. Taken together, the results demonstrate that people who claim to be sickened by electromagnetic fields cannot detect if the fields are present or not. A recent study at the University of Essex measured not only subjective feelings of well-being, but also objective parameters such as pulse, blood flow and skin conductance. No difference was found when the experiment was double-blind, but subjects who believed themselves to be electrosensitive reported lower levels of well-being when they were told that they were being exposed to electromagnetic fields.

Given the results of the double-blind, randomized trials and the lack of a plausible physiological explanation for electrosensitivity, what can we say about the unfortunate people who are convinced that their well-being is compromised by various electrical phenomena? That they suffer. That much is for sure. The question is, what are they suffering from?

MASCARA MASSACRE

COSMETICS ARE UNDER ATTACK. It's not the first time. Back in 1770, the English Parliament passed an act declaring that marriages could be pronounced null and void if the man had been "led into matrimony by false pretenses through the use of scents, paints, cosmetic washes, artificial teeth, false hair, bolstered hips, high heels or iron stays." It isn't quite clear what iron stays were, but the reference was probably to devices that steadied those features of the female anatomy which sometimes have a tendency to droop.

Whether any men actually sought a divorce based on their disappointment that the goods were not as advertised remains lost to history, but it is safe to assume that cosmetic manufacturers were not happy with the situation. They probably were not thrilled with Queen Victoria, either, when she publicly declared that makeup was improper, vulgar and acceptable for use only by actors. Today, cosmetics are being assaulted again, but for a different reason. They are being accused of harbouring potentially toxic ingredients, and regulatory authorities are being taken to task for not doing enough to protect the health of the public.

The finger-pointers range from faceless composers of inane emails to various activist organizations that bolster their crusade for "safer cosmetics" with references to the scientific literature. Some of the accusations, such as that certain mascaras or lipsticks contain toxic amounts of lead, are pure nonsense and can be dismissed out of hand. But allegations that certain cosmetics may contain hidden carcinogens or hormone-disrupting substances merit scrutiny.

The personal-care product industry is huge, netting some 250 billion dollars a year in global retail sales. Unlike pharmaceuticals, no pre-market testing of the safety of cosmetics is required, a fact often vociferously pointed out by cosmetic critics who infer that such lack of regulations puts consumers' health at risk. Actually, governments don't exactly maintain a hands-off policy. Canada has a "hot list" of some five hundred chemicals that cannot be used in cosmetics, and before any item is marketed, its list of ingredients has to be submitted to Health Canada for approval. Furthermore, Health Canada has the power to order the removal of products from stores if it decides there is any risk involved. Regulations are less stringent in the U.S., where the Food and Drug Administration has to prove that a product is dangerous before ordering it off the shelves.

One of the reasons that governments have not taken a heavy-handed approach towards requiring pre-market testing of cosmetics is that the industry has a very effective self-regulating program. The U.S.–based Cosmetic Ingredient Review panel is an industry-sponsored group of experts that includes representatives from the FDA as well as consumer organizations and is responsible for compiling and scrutinizing research that is relevant to cosmetic ingredients. The panel's in-depth reports are used by the industry to make decisions about product formulation. Oh, I can see some of you rolling your eyes right now at the mention of industry self-regulation. The fox is in charge of the henhouse, you may be thinking. Not so.

No industry wants to harm its customers. At the very least, including ingredients that turn out to be harmful is bad for business, especially in the U.S., with its litigious society and plethora of personal-injury lawyers advertising for prospective clients. Cosmetic companies know that the best way to make money is by selling products that are safe and effective. Admittedly, there is a fly in the ointment here, in that safety and efficacy are open to interpretation, as is the degree of acceptable risk associated with any consumer product. Yes, no matter how much care is taken, there is always some risk. Indeed, cosmetics can be responsible for some acute adverse effects, but these are usually readily recognized. It is the hypothetical links of some chemicals in cosmetics to cancer, or to endocrine disruptive effects, that are difficult to evaluate and provide grist for the alarmist mill.

The most serious acute effects include skin irritation and allergies. Sodium hydroxide in some hair straighteners or methacrylic acid in artificial nail products can be potent skin irritants if improperly used, and a host of chemicals ranging from fragrance components and preservatives to emulsifiers and colourants can cause allergic dermatitis. While the vast majority of consumers never encounter any such problems, a significant number do. An estimated fifty thousand people a year visit emergency rooms in North America for cosmetic-related problems, most of which turn out to be minor. Manufacturers of "hypoallergenic" cosmetics avoid the most obvious sensitizers such as lanolin, formaldehyde-releasing preservatives and fragrance components like cinnamic alcohol, geraniol, limonene or linalool.

Without a doubt, the scariest allegation is that cosmetics may contain carcinogens. Indeed, some do. It is important to realize, though, that the definition of a carcinogen is a substance that is capable of causing cancer in some animal at some dose. It does not mean that it is known to cause cancer in humans. Dioxane, for example, is found as an impurity in some cosmetics and is listed as a carcinogen because it triggers the disease when fed to rodents. Amounts in cosmetics, however, are vanishingly small, known to be present only because of the availability of extremely sensitive detection techniques. Furthermore, application of a chemical to the skin is not the same as ingestion. Still, elimination of any carcinogen is desirable, and methods to remove the dioxane impurity have been developed by major cosmetics manufacturers.

Cosmetic formulation is a continuously evolving process, one that keeps in step with the massive amount of research in the area. When new findings reveal a problem, the industry moves to address it. After all, consumer confidence is what puts money in the bank. Recent research linking parabens, commonly used preservatives, to the aging of skin cells, or some moisturizing creams with the promotion of skin cancer in mice after exposure to ultraviolet light merit further investigation, as does the possible hormonal effect of phthalates, chemicals used in some fragrances and nail polishes. While the relevance of these risks to humans is debatable, there is one major established cosmetic risk that is avoidable: applying mascara in a moving vehicle causes loads of eye injuries. Maybe we should consider passing a law to prohibit it.

TESTOSTERONE

ROOSTERS PROBABLY CONSIDERED Professor Arnold Berthold public enemy number one. But to scientists, he was a pioneer. And to athletes who abuse steroids, he's probably a hero— provided, of course, that they have heard of him. If they haven't, they should have. Because it was Berthold's classic experiments carried out at the University of Göttingen in Germany in 1849 that laid the foundations for research leading to the eventual isolation of testosterone, the main male sex hormone.

Professor Berthold was certainly familiar with the idea that removal of the testes caused profound changes in a man's behaviour. Eunuchs, for example, had long been known to have less aggressive personalities. Indeed, that's why Roman emperors like Constantine, fearing assassination, surrounded themselves with servants who had been rendered mild-mannered by removal of their manhood. (One suspects, though, that they weren't exactly mild-mannered during the procedure.) In any case, not only did castration affect behaviour, it also affected physiology. Eunuchs were less likely to go bald, and if castration took place before puberty, their boyish voices were retained. That's why in Italy, where women were forbidden to sing in church, *castrati* provided the high treble tones. The Sistine Chapel choir actually featured eunuchs into the early twentieth century.

The effect of castration on animals had also long been known. By 2000 BC, castration of male farm animals to make them easier to handle was widespread. Bulls, rams and stallions were all made more docile with a few well-targeted snips, rendering them less likely to protest when asked to haul loads or pull ploughs. But nobody really took much interest in just how castration brought about these dramatic changes until Berthold began his investigations. As it turned out, not ones that roosters would crow about.

Berthold happened to be curator of the local zoo and therefore had ready access to animals for experiments. Roosters were his choice. He found that when the birds were castrated, they stopped fighting with each other, lost their sex drive and no longer crowed. And there was a visible change as well: the roosters' combs became floppy. That was surprising, but not as much as the next phase of Berthold's experiment. He placed the testicles back into the roosters' body cavities and noted that normal male behaviour was soon restored and combs once again became erect! Realizing that all nerve connections had been severed when the testes were removed, Berthold correctly concluded that the testes influence behaviour by secreting some substance into the bloodstream. "The testes act upon the blood and the blood acts upon the whole organism," he famously stated. And then Berthold went one step further. He showed that the castration-induced changes were reversed by administering a crude testicular extract. It was the first example of hormone-replacement therapy.

Somewhat surprisingly, Berthold's work lay dormant for some forty years until it was resuscitated in 1889 by a startling article in the British medical journal *The Lancet*, with the enticing title "Note on the effects produced on man by subcutaneous injections of a liquid obtained from the testicles of animals." What made this paper truly remarkable was that the man in question was none other than the author, respected physiologist, Charles Brown-Sequard. To the astonishment of readers, the seventy-two-year-old Brown-Sequard reported dramatic rejuvenating effects after self-administering testicular extracts of dogs and guinea pigs. Why these particular creatures were selected as organ donors was not explained, but the difficulty of procuring human volunteers probably had something to do with it.

In any case, Brown-Sequard was obviously impressed with his handiwork: "A radical change took place in me . . . I had regained at least all the strength I possessed a good many years ago . . . with regard to the facility of intellectual labour, which had diminished within the last few years, a return to my previous ordinary condition became quite manifest." Furthermore, he went on to report, "a great improvement with regard to the expulsion of fecal matter." Well, speaking of the expulsion of fecal matter, perhaps Brown-Sequard's claims require a little scientific scrutiny.

Could there have been enough testosterone in the extracts to produce the changes he noted? Actually, while testosterone is produced in the testes, it is quickly passed into the circulatory system with very little being stored in the testes. Still, the only meaningful way to test the validity of Brown-Sequard's report is to repeat the experiment. And that is just what Australian researchers decided to do. Well, not exactly—they weren't foolhardy enough to inject themselves with testicular extracts, but they did prepare them exactly as Brown-Sequard described. And then, using standard laboratory techniques, they determined the amount of testosterone present. The results clearly indicated that Brown-Sequard could not possibly have benefited from any physiological effect. The amount of testosterone found was four orders of magnitude less than what is required to raise testosterone levels to normal in men who have lost their testes. Clearly, even though Brown-Sequard was an experienced researcher, he had not controlled for the placebo effect.

Newspapers reported widely on the renowned physiologist's "groundbreaking" work, and "organotherapy" fever gripped aging European men who lined up to be injected with the glandular extracts of diverse animals. Goat glands, because of the somewhat promiscuous nature of their owners, were in great demand. But most men found that whatever initial benefit they derived waned quickly, as is often the case with placebos, and organotherapy fell into disrepute.

Finally, when a few milligrams of pure testosterone were isolated from mountains of bull testicles in the 1930s, serious research on the hormone could begin. Unfortunately, proper controlled trials have failed to show any dramatic results with testosterone supplementation in aging men, save those who have lost testicular function. On the other hand, in larger doses, athletes have reaped the strength-enhancing benefits of testosterone, but at the cost of mortgaging long-term health. Testosterone really is a "Jekyll and Hyde" substance. Which is appropriate terminology, given that Brown-Séquard was Robert Louis Stevenson's neighbour in London, and was the author's inspiration for *Dr. Jekyll and Mr. Hyde.*

ALLERGIC TO IT ALL

THESE POOR SOULS SUFFER—that much is for sure. But just what are they suffering from? A whiff of perfume, the scent of paint, the smell of diesel fumes, the fragrance of a cleaning agent or the aroma of newsprint sends them reeling. Symptoms usually include headache, fatigue, breathing problems, muscle and joint pain, irritability, insomnia, short-term memory loss and confusion. How can such chemically diverse substances cause such similar symptoms? The explanation sometimes offered is that these unfortunate people suffer from something called "multiple chemical sensitivity (MCS)"—formerly "twentieth-century disease" and also referred to as "environmental illness" and "total allergy syndrome."

The theory is that an initial exposure to a chemical somehow throws the immune system into hyperactivity, causing it to become sensitive to tiny amounts of chemicals that are harmless to most people. Sounds sort of plausible, but most allergists and immunologists don't buy this explanation. Standard immune function tests show no abnormalities, they explain, and there is no known mechanism that can account for small doses of chemicals with completely different molecular structures simultaneously affecting so many body systems. They therefore raise the possibility that the symptoms may originate in the patient's mind, triggered by the stresses of everyday life. Stress, they suggest, can cause a variety of symptoms, and chemicals serve as a convenient scapegoat. People become sensitized not by the chemicals themselves, but by press reports that constantly raise the possibility of a link between illness and environmental contaminants. "No, no," counter a small group of practitioners, including some physicians, osteopaths and naturopaths who argue that multiple chemical sensitivity is a very real disease caused by the failure of some people's immune systems to adapt to the numerous synthetic chemicals that have been introduced into our environment since the Second World War. These "clinical ecologists," as they often call themselves, use a variety of techniques not accepted by orthodox medicine to diagnose and treat patients with multiple chemical sensitivity.

So, is multiple chemical sensitivity a psychological or a physiological disorder? Evidence would suggest that it can be either one. Personality profiles of MCS patients often show a history of psychological problems, depression and anxiety. Some see themselves as prisoners in a hostile world surrounded by chemicals that make them ill and leave them at the mercy of uncaring physicians who try to palm them off as psychiatric cases. Relief is to be found in retreating from modern life, sometimes by moving to remote communities and living in houses with walls lined with aluminum foil and furnished with items that release no smells of any kind. They drink only ultra-pure water, eat organic foods, scorn synthetic fibres and wear masks when they venture outside to protect themselves from the perfumes, pesticides and gasoline vapours they say make them sick. Some have to hang newspapers on a clothesline for hours to allow the smell of the ink to dissipate before bringing the paper into the house. Often these people will have a very poor understanding of physiological processes and sometimes come up with bizarre explanations for their affliction. One woman, for example, claims that her body's electromagnetic polarity runs counterclockwise instead of clockwise. Of course, even if the symptoms of MCS are triggered by the anxieties and frustrations of modern life, they are very real. The question is what to do to ease these patients' suffering.

Sometimes, clinical ecologists use a technique called provocation-neutralization. Small amounts of a suspect material are injected under the skin, and if symptoms are provoked, a larger dose is injected to "neutralize" the symptoms. The patient usually knows what is being injected and the power of suggestion can come into play. Lists of potentially offensive substances are compiled in this way, but studies show that when patients are challenged in a blinded test, they are unable to tell whether the injection contains an offending substance or not. It seems that belief in whatever treatment is used, and not the treatment itself, is the criterion for relief.

It would be a mistake, though, to suggest that all cases of MCS stem from an irrational fear of chemicals. Some chemicals, even in small doses, can cause a barrage of symptoms in some unfortunate people. And they don't have to be chemophobics! The case of a pharmacology professor, certainly not someone averse to working with chemicals, makes the point. He began to develop itching and irritation around the face when he worked at his desk. Over a period of a few weeks, the situation became worse: his eyes constantly watered, he was tormented by burning and stinging sensations and sometimes even respiratory problems. One day he became aware that *The New York Times* he was reading had an odour that disturbed him and realized that his problems could be related to paper. Indeed, his symptoms were exacerbated when he was near any paper product. Emptying his office of most paper helped, but he then started to react to felt-tip pens, gasoline odours, aftershave and even new permanent-press pants. Eventually he learned that just prior to the onset of his problems, the ventilation system in his office had been turned off for cleaning and had not been turned on again. It seems that the accumulation of certain compounds in the air, perhaps chemicals such as formaldehyde used in paper processing, had indeed wreaked havoc with his immune system. A classic case of multiple chemical sensitivity, and not one of psychogenic origin.

Without a doubt, MCS is a miserable affliction, no matter what the cause. Sufferers sometimes claim that they are the canaries of this chemical age, signalling calamities that are destined to become more widespread. Critics say they are more like cuckoos. Both analogies are probably off track. A better comparison would be with the hooded pitohui of New Guinea. This bird harbours a potent neurotoxin in its skin and feathers that affords protection against predators. It is fascinating, and rare, but very real. Just like multiple chemical sensitivity.

VITAMIN B₆: THE REAL STORY

NO, THE GOVERNMENT is not going to stop you from growing a herb garden. No, mothers will not require a licence to give vitamins to their children. No, inspectors will not go into private homes to seize dietary supplements. No, the government is not going to stop health food stores from selling natural health products. And no, Big Pharma is not trying to outlaw the sales of vitamin B$_6$ so that it can sell its own version of the vitamin as a prescription drug at an outlandish price.

These are just some of the totally false notions being spread by natural-health-product advocates in response to attempts by Health Canada and the Food and Drug Administration in the U.S. to tighten regulations governing the sale of such substances to the public. If you listen to some of the allegations, you would think that these agencies are proposing to tar and feather anyone who is caught trying to buy vitamins. Nonsense. All that Health Canada and the FDA strive for is a reasonable demonstration that natural health products can actually do what they claim, that they have an acceptable safety profile, that what is on the label is actually in the product and that all such products are produced according to the principles of good manufacturing practice. Sounds like motherhood and apple pie to me. Why should there be any objection to these regulations? Because there are numerous supplements on the market that make scientifically unfounded claims and stand to lose the profits they now rake in.

One of the common tactics used to boost the image and sales of natural health products is the charge that Big Pharma is trying to ban nutritional supplements because these wonderful products keep people from getting sick and therefore cut into the sales of expensive but ineffective prescription drugs. A typical attack comes from Mike Adams, a character with no formal scientific education, who has anointed himself with the ridiculous name of Health Ranger. Adams, who unjustifiably manages to garner a great deal of attention on the Internet, alleges that if governments are "allowed to keep banning nutritional supplements while promoting the very same drugs synthesized from those natural sources, it could allow Big Pharma to commit widespread biopiracy, stealing all the good medicine from nature, claiming patent protection on the useful molecules."

What prompted this silly bit of prattle? A petition submitted to the FDA by Medicure Pharma, a Canadian pharmaceutical company, asking the agency to prohibit the marketing of dietary supplements containing pyridoxal-5'-phosphate. Adams, along with other anti-pharma advocates, claims this means that Medicure is aiming to ban the sale of vitamin B_6 to the public. Poppycock!

Here is the real story. Vitamin B_6 is an essential nutrient required for the production of hemoglobin, the proper functioning of the immune system, the maintenance of blood glucose levels and myriad other important biochemical reactions. Our bodies cannot make vitamin B_6, so we need to ingest it in our diet. Since the vitamin is found in cereals, meat, fish, fruits and vegetables, deficiency in North America is extremely unlikely, but given that the vitamin has so many biochemical functions, a number of studies have been carried out to explore its potential in supplemental doses. These have included its possible use in depression, premenstrual syndrome, Parkinson's disease, carpal tunnel syndrome, chronic pain, autism and heart disease. While there have been some indications of potential benefit, the case for supplementation is less than compelling. Still, many consumers choose to take vitamin B_6, believing it may somehow be helpful. And nobody is against this, including Medicure Pharma, a company that believes in the therapeutic value of B_6. In fact, that is what led the company to investigate how the vitamin functions in the body and how it could be used to advantage.

After ingestion, vitamin B_6 is converted to pyridoxal-5'-phosphate (PLP), which is its active form. It is this substance that piqued the curiosity of Medicure researchers because laboratory investigations had suggested that it might curb reperfusion injury, a serious problem that patients experience when blood flow is restored to the coronary arteries after a heart attack or after coronary bypass surgery. The influx of fresh oxygenated blood can trigger the formation of free radicals that injure the walls of the artery, possibly precipitating a further cardiac event. Medicure invested millions of dollars in studies of patients who had undergone cardiac bypass surgery to investigate whether PLP, in doses far greater than that available from the diet, reduced the risk of reperfusion injury. In such doses, PLP functions as a drug, and given that vitamin B_6 is itself known to cause neurological problems in high doses, requires extensive study for possible side effects.

During the course of these investigations, Medicure learned that PLP was being sold to the public by some companies as a natural health product without any studies to back its use. While vitamin B_6 of course occurs naturally, its metabolite, PLP, does not. So how can it qualify as natural? The argument is that since it is found in the human body, it is natural. Well, if we go down that road, then why not sell estrogen, testosterone or cortisone as natural health products, since these are also found in the body? Obviously, that would be a risky business. Furthermore, none of the promoters of PLP have submitted applications to the FDA for its sale as a dietary supplement, as required by law.

Yes, Medicure is trying to protect its investment in research, but to allege that the company is trying to block the sale of vitamin B_6 is total nonsense. PLP is not the same as vitamin B_6, and all that Medicure is saying is that if PLP is to be used as a drug, it should be subjected to the same regulations as any other drug. The company has absolutely no desire to prevent the sale of vitamin B_6, a substance it believes to be integral to good health. Mike Adams's accusation that Medicure's petition to the FDA to stop the illegal sale of PLP is an example of Big Pharma trying to prevent the sale of vitamins is pure folly. But I guess we shouldn't be surprised at such comments, since Adams also tells people never to visit Western doctors, to ignore the "ridiculous" Food Guide Pyramid and to get an hour of full exposure to sunlight without any sunscreen, something he says is important for "vibrational nutrition" and mental health. Hmm. Maybe Adams hasn't been exposing himself to enough sun.

MY WHOLESOME, CARCINOGENIC SOUP

I LIKE TO LISTEN to the radio when I cook. And I was just in the midst of chopping veggies when the lead item on the newscast caused my ears to perk up: "Carcinogenic chemicals found in baby shampoos!" I knew I had better get ready for the onslaught of phone calls and emails because the words "carcinogen" and "baby" in the same sentence add up to a formula for panic. But first things first. The chicken was already in the pot and I had to finish chopping my parsnips, carrots, onions and celery.

Once the soup was nicely simmering, I began to Google. I learned that the consumer group Campaign for Safe Cosmetics had commissioned a laboratory analysis of a number of children's shampoos and bath products and was now trumpeting the discovery of undeclared formaldehyde and dioxane. Since both of these chemicals are classified as probable human carcinogens, it comes as no surprise that the airwaves and newspapers were soon filled with stories about baby products "tainted" with cancer-causing chemicals. What did I make of it all? Much ado about not much.

How do these chemicals end up in these consumer items in the first place? Formaldehyde is an indirect additive, released at a controlled rate from imidazolinyl urea, DMDM hydantoin or quaternium 15. Its purpose is to keep bacteria at bay. Dioxane is another story. This is a trace contaminant that is formed during the manufacture of certain detergents, such as those that may be included in baby shampoos and bubble baths. Concern arises because, in addition to being likely carcinogens, formaldehyde and dioxane are also potential allergens. Of course, the pertinent question is whether or not the amounts found in these products present a risk. As we well know, dose matters.

As far as allergens are concerned, the dose that matters can indeed be very small. Some cosmetic products may contain up to 600 parts per million of formaldehyde, and that can cause skin irritation. Such reactions are rare but possible. But reactions to bacterial contaminants that may form in the absence of preservatives are a bigger concern. It always comes down to a risk-benefit analysis. We don't ban peanuts because they threaten the lives of some who are allergic to them.

People may accept the risk of an allergic reaction, but when the spectre of cancer rears its head, that's another story. Formaldehyde *can* cause cancer, at least in animals that inhale it at a high dose over a long period. There is also some evidence that embalmers, pathologists and people who work with formaldehyde-containing resins may have a slightly increased risk of cancers of the lungs, nose and throat. But these exposures are astronomically greater than that presented by products such as shampoos. Even when formaldehyde is applied to the skin of mice at concentrations of up to 10 per cent over their lifetime, no effects on longevity are noted.

Dioxane, based on animal studies, is a carcinogen. When administered to rats in large amounts in their drinking water, it can cause cancer. However, studies carried out on people who have extensive occupational exposure to dioxane have not shown any increase in cancer rates. So why should we tremble at the trace amounts of dioxane in some cosmetic product, amounts which in any case evaporate almost immediately when applied to the skin? We shouldn't. But what about the possibility that tiny amounts may be absorbed through the skin and build up in the body? That seems to be unlikely. A study of more than two thousand people of all ages, randomly selected, has failed to detect any dioxane in their blood. So what's our bottom line? Exposure to trace amounts of formaldehyde or dioxane from cosmetic products is not a big worry. If we're going to get all antsy about such things, we might as well worry about—well, chicken soup.

My chicken soup is loaded with carcinogens. Not manufactured additives but naturally occurring compounds. Furocoumarins like 8-methoxypsoralen are present in parsnips and celery. Not only are they potential carcinogens, they can cause nasty skin reactions. Carrots contain caffeic acid, another carcinogen. I commonly add basil, which contains estragole, a known rodent carcinogen. The same can be said for alpha-methylpyrroline in black pepper. I'm sure an analysis of my cooked chicken would reveal some hetero-cyclic aromatic amines—nasty carcinogens. And then there's the formaldehyde. It occurs naturally in the onions and shiitake mush-rooms I use to flavour the soup. In fact, shiitake mushrooms can have a whopping 400 parts per million of formaldehyde.

What is my point here? It isn't to trigger headlines about toxic chicken soup. We need to realize that we are constantly exposed to thousands and thousands of chemicals, both natural and synthetic, every day. Some have the ability to trigger cancer under certain conditions. But the nature of those conditions is critical. The risk to an embalmer working with a concentrated formaldehyde solu-tion day in and day out, or to a rodent reared on formaldehyde-laced drinking water, is not the same as that to a human exposed to traces of formaldehyde in a shampoo. And I don't worry one bit about the formaldehyde I'm ingesting in my soup, even though I bet my exposure is far greater than from any shampoo.

If these notions drive you to drink, remember that alcohol is an established carcinogen. And if you would rather relax with a cup of coffee, well, enjoy the acrylamide, chlorogenic acid and furfural, along with a host of other natural carcinogens.

LIMITS OF ANIMAL TESTING

YOU'VE SEEN THE HEADLINES. Of the eighty thousand or so synthetic chemicals in the marketplace, only a few thousand have been adequately tested. And as far as the others are concerned, activist groups grumble that we are the guinea pigs that will determine their safety. According to them, we are all part of a massive, uncontrolled experiment, the consequences of which may turn out to be dire. No chemical should be introduced until it has been proven to be safe, they say. Sounds admirable, but just how does one prove safety? Testing for chemical safety is a very complex, time-consuming and, unfortunately, often unreliable process. And testing individual chemicals may not present a realistic scenario. For example, when mice are exposed in the womb to bisphenol A, the plastic component that is causing quite a commotion these days, they have a greater risk of developing diabetes, obesity and cancer. But when the pregnant mothers are given the B vitamin folic acid, or genistein, a compound found in soy, the effects of bisphenol A are negated.

What does this type of information mean for us? Hard to say. Obviously, we cannot do toxicity studies on humans, so we are left to scrutinize the results of animal studies and make educated guesses about the effects on people. Why guesses? Because a human is not a giant rat, a dog or a chimp. There are plenty of examples of substances that have appeared to be safe in animals and turned out to be toxic in humans, and vice versa. It wasn't long ago that six men in England who volunteered for an experiment to test a drug designed to treat diseases such as rheumatoid arthritis, leukemia and multiple sclerosis by dampening the body's immune reaction ended up in hospital, some suffering permanent organ damage. Mice, rats, rabbits and monkeys had shown no ill effects at all. But the men certainly did. As one of the volunteers who luckily had been given a placebo described, "The men went down like dominoes. They began tearing their shirts off complaining of fever, then some screamed that their heads were going to explode. After that they started fainting, vomiting and writhing around their beds." And this from a drug that had been shown to be "safe" in animals.

There are also cases of substances that cause problems in animals, but not in humans. If we used dogs as the standard animal to test food components, we could say goodbye to chocolate, a delicacy that is highly poisonous to dogs. Twenty-five grams of chocolate, a quarter of a chocolate bar, can supposedly kill a dog within a few hours. The culprits in the chocolate are compounds in the methylxanthine family, namely theobromine, theophylline and caffeine. In humans, they just deliver a small kick before they're metabolized by our liver enzymes. But dogs don't produce the same set of liver enzymes as we do, and the breakdown of the methylxanthines takes a much longer time. As these compounds circulate in the bloodstream, they affect the heart, the central nervous system and the kidneys. Unsweetened baking chocolate contains the highest concentration of these compounds, ten times as much as milk chocolate. So keep your dogs away from your chocolate cake. Even garden mulch made of cocoa bean shells presents a danger, although most dogs will not eat it.

Viagra raises another interesting point. When it was tested in beagles, it caused severe stiffness—not where it counted but in the neck. Researchers referred to this as "beagle pain syndrome." They also found that Viagra constipated mice and caused the livers of rats to swell. These problems were judged not to be severe enough to preclude human testing, and indeed it turned out that these side effects were not seen in men who took the drug. These are not unusual cases. A survey of some 150 compounds, which were produced by various pharmaceutical companies as prospective drugs but were never marketed because of some sort of toxic effect in people, revealed that only 43 per cent of these drugs caused similar problems in rodents and only 63 per cent did so in other animals.

The scientific literature is also full of examples of promising findings in laboratory animals that have not translated into effective treatments in humans. For example, tramiprosate (Alzhemed) was very effective in reducing the accumulation of amyloid protein in the brains of mice, a hallmark of Alzheimer's disease. Yet, as we saw previously, it failed in human clinical trials. So did statins, which showed promise in mice but have turned out to be ineffective in treating Alzheimer's patients.

Even closely related animal species do not necessarily respond the same way to chemicals. Take dioxins, for example. These are commonly described as the most toxic substances ever created. And they may well be—if you're a guinea pig. But the lethal dose for a hamster is one thousand times greater than that for a guinea pig. And we just don't know where humans fit into the scheme of things. Viktor Yushchenko, the Ukrainian president, was poisoned by a large amount of dioxin that somehow had been introduced into his food. Based on animal data, he should have died, but the only acute symptoms he suffered were inflammation of the liver and pancreas, along with facial palsy and a flare-up of a herpes infection. These quickly subsided, but the chloracne that characterizes dioxin poisoning did disfigure his face. There is also the possibility, based on animal data, that he is at risk for a type of cancer known as soft-tissue carcinoma. Certainly, it will be interesting for scientists to follow his progress.

Obviously, better models of testing are needed. And these could eventually come from testing chemicals on human cells in the laboratory. Of course, cells don't represent the whole organism, so there are still many issues here, but there is optimism. Techniques are being developed whereby liver or skin cells can be placed in thousands of tiny wells on a single dish, and different doses of chemicals can be systematically applied and the effects on the cells noted. Researchers are working on correlating results from such experiments with animal and human data, and within a few years we may in fact be able to test those thousands of chemicals to which we are exposed in a more reliable fashion. The next coal-mine canary may very well be an isolated liver cell in a laboratory dish.

AFTERWORD:
MAKING SENSE OF IT ALL

Why does a shower curtain get drawn in when taking a shower? How do two-in-one shampoos work? If nothing sticks to Teflon, how do they get it to stick to the pan? These are the typical kinds of questions I used to get on my radio show or after my public lectures. But my, oh my, how times have changed. These days, a question about shower curtains is likely to be about the release of phthalates; with shampoos, about the presence of parabens; and as far as non-stick cookware goes, I'm more likely to get sticky questions about health risks than cooking properties.

In general, questions tend to be in the "How much should I worry about such-and-such?" category. Depending on what's been in the news, the concern may be over fire retardants in upholstery, drug residues in drinking water, formaldehyde in bras, diiso-cyanates in mattresses, bisphenol A in canned food, trichloroethyl-ene in groundwater, dioxins in meat, mercury in vaccines, pesticide residues in food or radiation emanating from granite countertops. Curiosity about what we can do with chemicals has been replaced by fear of what chemicals can do to us.

All of this is very understandable, given that loss of health is our greatest fear in life. So, what can we do to prevent illness? Lifestyle factors such as proper nutrition, weight control and exercise are important. We also know that radiation, cigarette smoke and occupational exposure to certain chemicals can impact on health. But until recently, not much attention has been paid to exposure to the tiny amounts of chemicals that show up in the environment as a result of chemical innovations introduced since the end of the Second World War. Thousands of chemicals that never existed before are now produced in dazzling amounts and sometimes show up in unexpected places—like our bodies. Almost daily, media reports alert us not only to the presence of these chemicals, but to their potential for undermining our health.

Why is there so much interest in these environmental contaminants of late? First of all, we used to have bigger fish to fry. When you are concerned about improving food production, controlling malaria or battling infections, you tend not to sweat the small stuff—especially if you don't even know that the small stuff is there. Now, thanks to recent advances in technology, we know. Of course, it was always reasonable to suspect that our massive chemical production must leave some sort of environmental footprint, but we didn't worry much because of our reliance on the mantra of toxicology, namely that "only the dose makes the poison." We assumed that if we couldn't detect it, it couldn't be doing any harm.

Times have changed. Chemicals can now be detected at levels of parts per trillion, or sometimes even lower. And we have accumulating evidence that some, especially those that can mimic the effect of hormones, can produce physiological effects at such incredibly small concentrations. There is something else that we now have: much better information about disease incidence and disease patterns. Rates of childhood cancer have increased since the 1950s, as have prostate and breast cancers, both of which have a hormonal connection. Some of the increase undoubtedly can be attributed to better diagnostic techniques, but something else seems to be going on as well. The question is what. And there are experts aplenty who claim to know the answer. But the problem is that they all have different answers.

Research these days has become very specialized. Scientists who study the effects of bisphenol A released from plastics, for example, may know nothing about the work being done on phthalates or brominated diphenyl ethers or beryllium or electromagnetic radiation. Indeed, they may not even be aware of the existence of these fields of research. They live in separate worlds, the only common feature being the presence in each of these worlds of scientists, physicians or self-proclaimed experts who claim that our health problems are caused by their pet culprit. You can take your pick from plasticizers, dioxins, chlorates, air particulates, perfluorooctanoates or a slew of others. Since we are exposed to most of these to some extent, if the claims of harm were all true, we would be dropping like flies.

This is not to suggest that such environmental contaminants cannot have an effect on our health. They almost surely can. But teasing out which ones, and under what conditions they may cause harm, is a daunting task with many possible pitfalls. For example, a type of rat known as the Sprague Dawley is commonly used to evaluate compounds that may either mimic or block the activity of natural hormones. This rodent, though, has been bred to reproduce in a prolific fashion and has a hormonal constitution that may be more resistant to endocrine-disrupting chemicals. Probably not a good model for humans.

Any alteration in the genes that make up DNA upon exposure to a chemical is also a basis for determining toxicity, but we now know that certain chemicals can significantly affect the functioning of DNA without altering its structure. The burgeoning field of epigenetics deals with the notion of chemicals turning genes on or off without affecting their structure. A chemical, such as bisphenol A, for example, may deactivate a gene that codes for the production of a protein that helps protect a cell from cancer. Epigenetic research will probably be able to help focus our worries.

Cherry-picking data, a common process for people or organizations with agendas, is not the way to do science. Environmentalists, industry representatives, activists of all sorts and even government officials are in on the cherry harvest, selecting data that support a particular position, usually a controversial one, while ignoring relevant contradictory evidence. Of course, these days there is no shortage of scientific controversies. Concerns over the safety of plastic components, cosmetic ingredients, medications, pesticides, genetically modified organisms, cell phones, microwave ovens, drinking water and food are very much in the public eye. People search for guidance in tackling these issues, and they look for dispassionate, unbiased answers. In theory, that is just what scientific research should be able to deliver.

In the best of all possible worlds, scientists would all be competent, would have no preconceived biases, would not be driven by monetary gain, would have access to plentiful funding from unbiased sources and would have their egos safely tucked away. Alas, we do not live in Utopia. Indeed, the only uncontestable remark that can be made about current scientific research is that it is plentiful. Actually, that is an understatement—its sheer volume is mind-boggling. Thousands of peer-reviewed research papers are published every week, and they are obviously not all of equal quality. Contradictory findings are not uncommon, even in the absence of any bias. As a result, scientific publications can be found to support almost any point of view. You just have to be prepared to do a little cherry-picking.

If you want to "prove" that the pesticide DDT was responsible for an increase in breast cancer on New York's Long Island, you can certainly find peer-reviewed publications to back up the claim. But you can also "prove" the opposite by pointing at a seven-year federally funded study that showed no such effect. Studies can be found linking bisphenol A with developmental problems in rodents, but the literature also reveals plenty of studies absolving the chemical of blame. Antioxidant supplements are effective in warding off disease according to some studies, and useless according to others. This doesn't mean that any of the studies are wrong; it just highlights the difficulty of obtaining conclusive results when many variables are involved. That's just the nature of science.

But science gets short-shifted and the public ends up being misled when data are presented selectively to support an agenda. And when the stakes are high, this is almost inevitable. Unfortunately, at McGill's Office for Science and Society, we witness this on a daily basis as we try to make heads or tails of scientific controversies. Some public-spirited group will get on its high horse and claim, with supporting studies, that one chemical or another in our environment is wreaking havoc with our health. Industry then responds by circling its wagons and trotting out contradictory studies. Both sides sound convincing. Governments get involved, but any action has to take both activist and corporate sensitivities into account. There is name-calling aplenty on all sides, with the combatants attempting to outmuscle each other by enlisting reputable scientists to further their causes. These scientists may be hired directly by the organizations involved, they may have connections through sponsored research, or they may just support a certain view out of conviction. Each of these scenarios raises concerns.

Working directly for a vested interest is not conducive to objectivity. There is plenty of historical evidence for scientific philandering in, for example, the tobacco, asbestos, chemical, pharmaceutical and dietary supplement industries. Either the risks of the products are downplayed, or the benefits overstated, which can readily be done by referring to the literature selectively. Environmentalists have also been known to play the same game. And much too often, the response to an unfavourable study is to try to discredit the research by searching for flaws in the methodology instead of scrutinizing the work objectively.

With cutbacks to government-sponsored research, scientists are increasingly looking to industry for grants. Although it is nonsensical to dismiss a study just because it may have been sponsored, there is clear evidence that such studies are more likely to favour industry than studies that are totally independent. On occasion, even independent researchers may be so convinced of the merits of their pet view that they become self-delusional and ignore any contrary evidence.

Science should, of course, be objective and pristine. Let the data rule—*all* the data! But teasing out a sound scientific conclusion from the overwhelming amount of information available today is a daunting task. It requires detachment from any vested interest, expertise in evaluating the quality of studies and recognition of the fact that experimental results can be misinterpreted or purposefully twisted. Take, for example, the claim by a plastic bag manufacturer that bags responsibly disposed of in a landfill minimize global warming by performing a carbon-sequestration function. Hmm. The implication is that if the petroleum used to make the bag had instead been burned for fuel, it would have contributed to global warming. I suppose that is true, but it is playing a little loose with the facts. Talk about cherry-picking data!

So what is the point of all of this? That it is far more complicated to answer questions about health than about the behaviour of shower curtains. As for individuals who think their pet toxin is responsible for all of society's ills, well, they could use a cold shower. And for me, my worry is that I'm not sure what to worry about. But I don't think worrying about everything is the answer.

Our best bet is to accumulate as much information as we can before formulating an opinion. And that opinion is subject to change over time as more data accumulates. By and large science is a self-correcting discipline. Future research may absolve come chemicals of the crimes of which they have been accused or may lead to their ban. Today's sense may be tomorrow's nonsense.

In the case of controversial issues, all we can do is make educated guesses based on currently available information. But should contrary data emerge, there is no shame in changing one's opinion or course of action. Indeed, the proper pursuit of science demands it.

INDEX